Complexity: A Very Short Introduction

VERY SHORT INTRODUCTIONS are for anyone wanting a stimulating and accessible way in to a new subject. They are written by experts, and have been translated in to more than 40 different languages.

The Series began in 1995, and now covers a wide variety of topics in every discipline. The VSI library now contains over 350 volumes—a Very Short Introduction to everything from Psychology and Philosophy of Science to American History and Relativity—and continues to grow in every subject area.

Very Short Introductions available now:

THE U.S. CONGRESS Donald A. Ritchie
THE U.S. SUPREME COURT
   Linda Greenhouse
UTOPIANISM Lyman Tower Sargent
THE VIKINGS Julian Richards
VIRUSES Dorothy H. Crawford
WITCHCRAFT Malcolm Gaskill

WITTGENSTEIN A. C. Grayling
WORK Stephen Fineman
WORLD MUSIC Philip Bohlman
THE WORLD TRADE
   ORGANIZATION Amrita Narlikar
WRITING AND SCRIPT
   Andrew Robinson

Available soon:

HORMONES Martin Luck
GENES Jonathan Slack
GOD John Bowker

KNOWLEDGE Jennifer Nagel
CONFUCIANISM
   David K. Gardner

For more information visit our website

www.oup.com/vsi/

John H. Holland

COMPLEXITY

A Very Short Introduction

OXFORD
UNIVERSITY PRESS

OXFORD
UNIVERSITY PRESS

Great Clarendon Street, Oxford, OX2 6DP,
United Kingdom

Oxford University Press is a department of the University of Oxford.
It furthers the University's objective of excellence in research, scholarship,
and education by publishing worldwide. Oxford is a registered trade mark of
Oxford University Press in the UK and in certain other countries

Published in the United States of America by Oxford University Press
198 Madison Avenue, New York, NY 10016, United States of America

British Library Cataloguing in Publication Data
Data available

Library of Congress Control Number: 2014931680

ISBN 978-0-19-966254-8

Printed and bound by
CPI Group (UK) Ltd, Croydon, CR0 4YY

Preface

It came as a surprise when Oxford University Press approached me to write a *Very Short Introduction* on complexity, and at first I was quite reluctant. I had just finished a book on complexity, aimed at a science-oriented audience, and revisiting the topic so soon was not appealing.

Two properties of the VSI series changed my mind: The book was to be aimed at a much broader audience—anyone, of whatever background, interested in complexity—and it was to be short. In addition, my colleagues unstintingly praise the whole VSI series, touting the books as providing easy, authoritative overviews to any topic in its long list. All in all, it was a challenge that was both daunting and intriguing. I decided to see what I could do.

A bit to my surprise, the process of writing this book substantially increased my own comprehension, even though I had been teaching courses on complexity, at various levels, for many years. As the final chapter of this book began to take form, I saw that topics I had previously taken to be well separated actually had much in common. Reducing the concepts to their simple underpinnings served to 'glue' them together. I can only hope that you, the reader, feel the same.

John Holland
Ann Arbor, Michigan
November 2013

Contents

List of illustrations

Chapter 1
Complex systems

What is complexity?

Complexity, once an ordinary noun describing objects with many interconnected parts, now designates a scientific field with many branches. A tropical rainforest provides a prime example of a *complex system*. The rainforest contains an almost endless variety of species—one can walk a hundred paces without seeing the same species of tree twice, and a single tree may host over a thousand distinct species of insects. The interactions between these species range from extreme generalists ('army' ants will consume most anything living in their path) to extreme specialists (Darwin's 'comet orchid', with a foot-long nectar tube, can only be pollinated by a particular moth with a foot-long proboscis—neither would survive without the other). Adaptation in rainforests is an ongoing, relatively rapid process, continually yielding new interactions and new species (orchids, closely studied by Darwin, are the world's most rapidly evolving plant form). This lush, persistent variety is almost paradoxical because tropical rainforests develop on the poorest of soils—the rains quickly leach all nutrients into the nearest creek. What makes such variety possible?

Many other systems important to humans exhibit similar complexities: markets with their varieties of buyers and sellers, organized into groups participating in mutual funds; economies with hierarchies of workers, departments, firms, and industries; multi-celled organisms consisting of proteins, membranes, organelles, cells, and organs; the Internet with users, stations, servers, and websites; to name a few. Each of these complex systems exhibits a distinctive property called *emergence*, roughly described by the common phrase 'the action of the whole is more than the sum of the actions of the parts'.

In addition to complex systems, there is a subfield of computer science, called *computational complexity*, which concerns itself with the difficulty of solving different kinds of problems. The *travelling salesman problem* is a classic example: What is the shortest tour of a set of cities (say 20 cities) that visits each city once and only once? An attempt to solve this problem by exhaustively comparing all possible tours is enormously time-consuming (there are $19! = 19 \times 18 \times ... \times 1 > 10^{17}$ possible tours of 20 cities), and a small increase in the number of cities causes an inordinate increase in the search time (from $19!$ to $29! > 10^{31}$ in going from 20 to 30 cities). The object of the computational complexity subfield is to assign levels of difficulty—*levels of complexity*—to different collections of problems. There are intriguing conjectures about these levels of complexity, but an understanding of the theoretical framework requires a substantial background in theoretical computer science—enough to fill an entire book in this series. For this reason, and because computational complexity does not touch upon emergence, I will confine this book to *systems* and the ways in which they exhibit *emergence*.

Herbert Simon, in his classic book *The Sciences of the Artificial*, outlined the challenges posed by *complex systems*; he did this so well that, half a century later, the major points apply directly to current research:

> one might conjecture there has been a long-run trend toward variety and complexity....forms can proliferate because the more complex arise out of a combinatoric play upon the simpler....
> [E]volution is not to be understood as a series of tournaments for

the occupation of a fixed set of environmental niches.... Instead
evolution brings about a proliferation of niches.... Each new bird
or mammal provides a niche for one or more new kind of flea. (p. 189)

complex systems will evolve from simple systems much more
rapidly if there are stable intermediate forms.... The resulting
complex forms... will be hierarchic. Among possible complex
forms, hierarchies are the ones that have time to evolve.

In hierarchic systems we can distinguish between interaction
among subsystems... and the interactions *within* subsystems....
The interactions at different levels may be, and often will be, of
different orders of magnitude. (p. 209)

Much of my effort here will be to bring Simon's points into the
current, more nuanced setting. In every case, the objective is to
attain some ability to 'steer' the complex system. But how do we
steer a system through a complex, tangled web of interactions?

Complicated vs. complex

At the outset, it is helpful to distinguish *complex* from *complicated*.
The dictionary is not helpful in making this distinction, treating
the terms as almost synonymous and emphasizing 'interconnected
parts' in both cases. Moreover, distinguishing complex from
complicated involves the 'pile of sand' conundrum. If we start with
a recognizable pile of sand and start taking away one grain at a
time, when does it cease to be a pile of sand? At some point, the
pile of sand will disappear and most will see only scattered grains
of sand, but there is no sharp boundary at which this happens.
Complicated and *complex* similarly defy any attempt to provide a
sharp, distinguishing demarcation—it is easy to distinguish one
from the other at the extremes but there is a middle-ground where
the distinction becomes unclear and arbitrary.

For these reasons, complexity, like *life* and *consciousness*, does not
have a rigorous definition. However, as with life (biology) and

consciousness (psychology), this lack does not forestall a rigorous approach to the subject matter. Better yet, *emergence* ('the whole is more than the sum of the parts') helps distinguish complex systems from other systems. Indeed, historically, complexity became an increasingly important topic as physicists became intrigued with emergent properties of aggregates of identical elements, such as the 'wetness' of an aggregate of water molecules. There is no reasonable way to assign 'wetness' to individual molecules; wetness is an emergent property of the aggregate. In this, wetness differs from a property like weight, where the weight of the aggregate is simply the sum of the weights of the component parts.

Emergence itself is a property without a sharp demarcation. There are conflicting definitions, some claiming that emergence should be a holistic property, incapable of being reduced to the interaction of parts. That is not the interpretation used here. Instead this book concentrates on interactions where the aggregate exhibits properties *not* attained by summation. In mathematical terms, the interactions of interest are *non-linear*. This non-linearity, following Simon, yields levels of organization and hierarchies—selected aggregates at one level become 'building blocks' for emergent properties at a higher level, as when H_2O molecules become building blocks for water.

Hierarchical organization is thus closely tied to emergence. Each level of a hierarchy typically is governed by its own set of laws. For example, the laws of the periodic table govern the combination of hydrogen and oxygen to form H_2O molecules, while the laws of fluid flow (such as the Navier-Stokes equations) govern the behaviour of water. The laws of a new level must not violate the laws of earlier levels—that is, the laws at lower levels constrain the laws at higher levels. Any laws (or a theory) that violated laws at more elementary levels would typically be discarded. Restated for complex systems: emergent properties at any level must be consistent with interactions specified at the lower level(s).

Even within these constraints, there can be cases where emergence is questionable. A watch is well-described as a *complicated* train of gears and springs, but, at a stretch, we can think of 'time-keeping' as an emergent property of the watch. Is the watch, then, a *complex system*? Though we could say that a watch induces the user to 'wind it' when it 'runs down' (a top–down interpretation), that interpretation is not likely to greatly affect our understanding of watches, and it does not much help us in making comparisons with closely related systems (such as digital time-keepers). Much of the motivation for treating a system as complex is to get at questions that would otherwise remain inaccessible. Often the first steps in acquiring a deeper understanding are through comparisons of similar systems.

By treating hierarchical organization as *sine qua non* for complexity we focus on the interactions of emergent properties at various levels. The combination of 'top–down' effects (as when the daily market average affects actions of the buyers and sellers in an equities market) and 'bottom–up' effects (the interactions of the buyers and sellers determine the market average) is a pervasive feature of complex systems. The present exposition, then, centres on complex systems where emergence, and the reduction(s) involved, offer a key to new kinds of understanding.

The behaviours of complex systems

Complex systems exhibit several kinds of telltale behaviour. I will describe some of these behaviours briefly here; they will be examined in more detail in later chapters.

- *self-organization* into patterns, as occurs with flocks of birds or schools of fish
- *chaotic behaviour* where small changes in initial conditions ('the flapping of a butterfly's wings in Argentina') produce large later changes ('a hurricane in the Caribbean')

- *'fat-tailed' behaviour,* where rare events (e.g. mass extinctions and market crashes) occur much more often than would be predicted by a normal (bell-curve) distribution

- *adaptive interaction,* where interacting agents (as in markets or the Prisoner's Dilemma) modify their strategies in diverse ways as experience accumulates.

In addition, as already mentioned, *emergent behaviour* is an essential requirement for calling a system 'complex'.

Any careful study of these behaviours, especially when different complex systems are to be compared, requires techniques from several disciplines. Models, measures, and insights from the more traditional disciplines—physics, biology, computer science, economics, and mathematics—can be used, but they must be used in novel ways to discover 'laws' that hold across a wide range of complex systems.

Two kinds of complexity

As the field of complexity studies has developed, it has split into two subfields that examine two different kinds of emergence: the study of *complex physical systems* (CPS) and the study of *complex adaptive systems* (CAS):

The study of complex physical systems focuses on geometric (often lattice-like) arrays of elements, in which interactions typically depend only on effects propagated from nearest neighbours. An early tour de force in CPS studies was von Neumann's 1956 proof that a self-reproducing machine could be designed on a checkerboard-like array, called a *cellular automaton* (as discussed in Burks' *Essays on Cellular Automata*). Each square in the array can be in just one of a fixed, common set of 29 distinct states (in effect each square can contain one of 29 different particles); changes in state are determined by a law that is the same for every square, and looks only to the states of adjacent squares. That is, as with all

CPS, there is a 'universal law' and a geometry (think of 'gravity' and three-dimensional space, à la Newton).

To design his machine, von Neumann carefully selected a set of states corresponding to basic parts of the machine, much as one would selected a set of gears, shafts, and the like to construct a watch. He then arrayed copies of these parts, one to a square, over adjacent squares of the checkerboard, in effect connecting them to form the machine. The cellular automaton laws take over once the pattern is in place, causing the states to change without further input from the designer. Each state changes as dictated by the states of the eight surrounding squares, just as the different numbers of teeth in the gears in a gear-train fully determine their relative rates of rotation.

Part of von Neumann's insight was finding a pattern that, in due time, would reproduce itself in an adjacent set of empty squares. To make this a non-trivial task (as would be the case if, say, a single state were copied into an empty adjacent square) von Neumann required that the overall pattern be able to act as a general-purpose computer. Neither the possibility of a pattern acting as a general-purpose computer nor the possibility that a non-trivial pattern could reproduce itself was at all obvious at the start. However, von Neumann actually constructed a complex pattern—clearly not a living object—that was both capable of executing arbitrary algorithms and could demonstrably reproduce itself (Figure 1).

Prior to von Neumann's demonstration, it was held that only living things could reproduce themselves; von Neumann's 'existence proof' made it clear that machines could reproduce themselves as well, completely changing our definition of 'life'. As we will see, the study of CPS has a distinctive set of tools and questions centring on elements that have fixed properties—atoms, the squares of the cellular automaton, and the like. Under Nobel Laureate Phil Anderson's rubric 'more is different' (also discussed

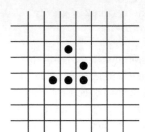

Laws ('Game of Life'):

A ball is created in an empty square if exactly 3 of the square's 8 neighbours contain a ball on the previous time-step; otherwise the square remains empty.

A ball persists in a square if 2 or 3 of its neighbours contain a ball; otherwise it is deleted and the square becomes empty.

All squares are updated simultaneously.

Evolution of a pattern ('glider') under the above laws:

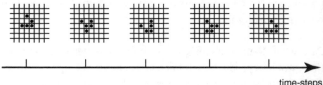

time-steps

Complexity

1. A cellular automaton

in Waldrop's classic, *Complexity*) the study of CPS has led to a better understanding of physical phenomena as widely different as Anderson's work on superconductivity, Turing's reaction/diffusion equations for pattern formation during morphogenesis, and the well-known 'butterfly effect'. The tools used for studying CPS come, with rare exceptions, from a well-developed part of mathematics, the theory of partial differential equations, the kinds of equations that Maxwell used to model electromagnetism, or that Navier-Stokes used to extend Newton's equations to studies of fluid flow. The nature of these approaches will be discussed in the next chapter, avoiding mathematical formalism wherever possible.

CAS studies, in contrast to CPS studies, concern themselves with elements that are not fixed. The elements, usually called *agents*, learn or adapt in response to interactions with other agents.

A commodities market provides a good example. The agents that do the buying and selling adapt their strategies as market conditions change, being strongly influenced by top-down aggregate averages, such as the market's price index. In the market there are continuing exchanges of information, in the form of buy and sell offers, and there are precipitous cascading effects such as the oft-observed 'bubbles' and 'crashes'. Solutions to some of the most important problems of the 21st century—enhancing the immune system, making ecosystems sustainable, regularizing global trade, curing mental disorders, encouraging innovation, and so on—depend upon a deep understanding of the interaction of adaptive agents in CAS.

It is unusual for CAS agents to converge, even momentarily, to a single 'optimal' strategy, or to an equilibrium. As the agents adapt to each other, new agents with new strategies usually emerge. Then each new agent offers opportunities for still further interactions, increasing the overall complexity. Tropical rainforests provide an excellent example. The heavy rainfall in the forest quickly leaches nutrients into the nearest stream—the soil in a tropical rainforest, as mentioned earlier, is one of the poorest soils in the world. In response to this shortage, agents in the forest act as 'basins' for retaining the nutrients. The bromeliad provides a case in point: it is a cup-like plant growing high in the trees, holding water and nutrients. Insects and amphibians lay eggs in the retained water, and the waste products of their larvae contribute to the bromeliad's nourishment. The interactions become ever more entangled as the amphibians eat the grown insects, and so on. All CAS exhibit similar entanglements, each new agent offering possibilities for still other agents. The complex feedback loops that form make it difficult to analyse, or even describe, CAS.

Analysing complexity

Analysis of complex systems almost always turns on finding recurring patterns in the system's ever-changing configurations.

The game of chess provides a useful analogy: a dozen rules determine the legal moves of a small number of pieces—32 pieces of 6 different kinds—arrayed on an 8x8 field of squares. Consider now the possibilities. If, on average, each move offers a choice of 10 legal possibilities, and an average game lasts 50 moves, then there are 10^{50} allowable move sequences—that's a number much larger than the estimated number of atoms in the universe. Indeed, in serious play, no two games of chess are ever identical. This *perpetual novelty*, produced with a limited number of rules or laws, is a characteristic of most complex systems: DNA consists of strings of the same four nucleotides, yet no two humans are exactly alike; the theorems of Euclidean geometry are based on just five axioms, yet new theorems are still being derived after two millennia; and so it is for the other complex systems.

How then are we to study systems that generate perpetual novelty using limited resources? Again, it is helpful to think about chess. Though chess games do not repeat, there are recurring patterns in the game, some so common that they have been given names: 'pin', 'fork', 'discovered check', etc.; books are written on how to use these patterns to steer the game to a win. As with chess, complex systems typically exhibit recurring patterns that offer similar possibilities for 'steering'. For example, in molecular genetics recurring patterns are called *motifs*, and in Euclidean geometry *derived rules* and *lemmas* help develop new theorems. To exploit these possibilities, then, analysis depends upon methods for discovering and exploiting recurring patterns in generated systems.

Two concepts from physics take centre-stage at this point: *laws* and *states*. Laws, like Newton's laws of gravity or Maxwell's laws of electromagnetism, are the counterpart of the rules of chess. States, such as the state of an economy, are the counterpart of particular configurations of pieces on the chessboard. As in chess, the laws determine the ways in which the states can change over time.

In *complex physical* systems, laws constrain the way in which a given initial state can change; in Newton's case, the state sequences (more formally, the *state trajectories*) range from the regular elliptical motion of the planets to the complicated trajectory of a planetary probe. As mentioned in the previous section, the laws of CPS are almost always formulated using *partial differential equations* (PDEs in standard physics parlance), where the variables of the equations specify the states (e.g., position, momentum, and time in Newton's equations). Even when the laws are precisely formulated in terms of PDEs, novelty is ever-present—it was decades before the radio waves implied by Maxwell's equations were actually generated.

In CAS the elements are adaptive agents, so the elements themselves change as the agents adapt. The analysis of such systems becomes much more difficult. In particular, the changing interactions between adaptive agents are not simply additive. This non-linearity rules out the direct use of PDEs in most cases (most of the well-developed parts of mathematics, including the theory of PDEs, are based on assumptions of additivity). The difficulty is compounded because in most disciplines involving CAS, such as the social sciences, there is no standard language for describing, let alone analysing, the interactions of agents. This lack is surprising because there are precise languages describing other kinds of complex human interactions, such as the scores used in music or the choreographic notation used in ballet. True, music and ballet are more stylized than the subtle, everyday social interactions that take place between humans, but aspects of those interactions can be stylized (as in stylized studies of cooperation using the two-person Prisoner's Dilemma game). To steer CAS we must go beyond collecting and organizing data to discover the mechanisms that generate the data, and to do that we require a precise language for describing the adaptive interactions of large numbers of agents.

There is a suggestive precedent in language studies: the *universal grammar* (UG) formalism, introduced by Noam Chomsky. A formal grammar à la Chomsky involves a set of *generators* (e.g. a vocabulary), and a set of *operators* for combining the generators into meaningful strings (e.g. sentences). The purpose of a formal grammar is to generate a *corpus* (set) that describes the states (sentences) that can occur under the grammar's rules. For both CPS and CAS there is a time dimension that must be incorporated, so that the order in which an element is generated by the grammar (the placement of a word in a sentence) is explicitly attached to that element (in a subject–verb–object grammar, a verb is ordinarily placed *after* a subject noun. In a broader context, the order of generation of elements in the corpus plays the role of time. The effect is much like setting up and executing a computer program, where the generators are the instruction set and the corpus is the result of executing the program.

This analogy to computer programming once again emphasizes the difficulty of using PDEs to understand generated systems. In a typical physical system the whole is (at least approximately) the sum of the parts, making the use of PDEs straightforward for a mathematician, but in a typical generated system the parts are put together in an interconnected, non-additive way. It *is* possible to write a concise set of partial differential equations to describe the basic elements of a computer, say an interconnected set of binary counters, but the existing theory of PDEs does little to increase our understanding of the circuits so-described. The formal grammar approach, in contrast, has already considerably increased our understanding of computer languages and programs. One of the major tasks of this book is to use a formal grammar to convert common features of complex systems into 'stylized facts' that can be examined carefully within the grammar. Along the way, we will see that generated systems have close relations to other techniques used in analysing complex systems, such as network theory and data-mining.

Chapter 2
Complex physical systems (CPS)

Characteristics of complex physical systems

The elements of a CPS follow fixed physical laws, usually expressed by differential equations—Newton's laws of gravity and Maxwell's laws of electromagnetism are cases in point. Neither the laws nor the elements change over time; only the positions of the elements change. Under such determinism it would seem that systems with similar starting points would unfold in similar ways—as with the regular orbits of planets or the propagation of light. And such is often the case. However, in the latter quarter of the 20th century, it became increasingly clear that there are important exceptions to this regularity. The fluid flow equations used in the prediction of weather provide a good example. It was discovered that minute differences in initial conditions for the weather equations could lead to later outcomes so different that they appear random relative to the starting point. Somewhat fancifully, the flap (or not) of a butterfly wing in the Pampas of Argentina can result in the appearance (or absence) of a hurricane in the Caribbean at a later point in time. The course over time appears chaotic rather than deterministic. Though the partial differential equations (PDEs) describing *chaotic systems* are fully deterministic, the presumed guarantees of determinism—similar starting conditions yield similar state trajectories—no longer hold. Nevertheless weather prediction is possible, a point we'll want to examine more closely later.

Per Bak, in the 1980s, set a distinctive tone to the study of CPS. His famous example examines the effect on a conical sandpile of a narrow stream of sand grains falling on its apex: at a critical stage, cascades of sand avalanche to the bottom of the pile, completely changing its conical shape. Bak named this phenomenon *self-organized criticality*. Current studies of CPS build on these ideas to investigate systems ranging from superconductive transmission of electric currents to quantum computers and messages where the sender and receiver are immediately aware of any interception (quantum encoding). These properties and others, such as self-similarity, scaling, and power laws, can only be touched upon in this short introduction, but each will be described.

Snowflake curves and fractals

Self-similarity is a good place to start, and the *snowflake curve* offers an easy example. The curve is constructed by repeated use of the same construction, hence the self-similarity. It starts with an equilateral triangle. Each side is divided into 3 equal parts and then an equilateral triangle, with sides one-third the original, is placed with its base on the middle third of each side of the original (see Figure 2). Then all interior lines are erased, resulting in a closed 'curve' made up of 12 equal-length straight segments and six 120-degree angles. This procedure is then repeated for each of the 12 new straight-line segments. By successive repeats, one gets a curve made up of more and more angles (3, 6, 12, 24, . . .) and ever more (3, 12, 48, . . .), and ever shorter, straight-line segments. The resulting curve looks a bit like the outline of a snowflake,

2. A snowflake curve

hence the name. As the straight segments get smaller and smaller we arrive at a curve that is mostly angles. More formally, in the limit, the curve is everywhere discontinuous.

Almost as soon as the snowflake curve was defined it became apparent that there were whole classes of snowflake-like curves, called *fractal curves*, generated by simple laws. Moreover, with these curves in mind, observers began to find natural systems exhibiting self-similarity: coastlines, tree branching, mammalian circulatory systems, and so on. These curves and systems posed a dilemma for the belief, widely held by 19th-century scientists, that the behaviours generated by simple laws (e.g., the *state trajectories* generated by physical laws) were smooth and continuous. Quantum theory, though ostensibly concerned with discontinuous changes, did not change this outlook: it simply replaced continuous trajectories with probabilistic trajectories (the *wave functions*). Formalisms that assume continuity, such as PDEs, do little to unveil the properties of self-similar systems. These departures from continuity put increased emphasis on new ways to describe and understand systems we now call 'complex physical systems'.

Scaling

Scaling is a property related to self-similarity, suggesting a new way to examine CPS and complexity in general. *Zipf's law*, derived from observations of word use in different languages, provides an easily understood example of scaling. Zipf's observation was that, approximately, the most frequently used word in a language (any language) will occur twice as often as the second most frequent word, three times as often as the third most frequent word, and so on. In mathematical terms, when words are ordered according to their usage (1st, 2nd, ...) and then plotted against the frequency of that usage, the resulting curve is exponential. Such laws are called *power laws*.

Thoughts about scaling originated in 19th-century biology through observations of the metabolism of animals of different sizes. The heat produced by an animal's metabolism is determined by the number of cells contained within its outer membrane, approximately proportional to the volume enclosed. However, the animal must dissipate this heat through that enclosing surface. Elementary geometry tells us that the surface area of, say, a sphere goes up as the square of the sphere's radius, while the volume enclosed by the spherical surface increases as the cube of its radius. Thus, to the extent an animal's enclosing membrane is sphere-like, the ability to dissipate the heat relative to the rate of heat production should change, as animal size increases, at a square/cube, or 2/3, power, rate. A large animal, then, because it has proportionately less area for dissipating heat should be *very* hot compared to small animals. But that is *not* the case. Why?

One solution to the heat problem would be for animals to have fractal surfaces with increasing numbers of indentations, so that the enclosing area increases rapidly relative to the volume. But animals do not have fractal surfaces. A different solution would be to decrease the metabolic rate in proportion to the surface/volume exponent—that is, the metabolic rate would decrease as the 2/3 power when plotted against animal size. This is a reasonable conjecture, and it was approximately confirmed by measurement. However, in the first third of the 20th century, more careful measurements of metabolic rates established that 3/4, instead of 2/3, more closely approximated the metabolic rate/size exponent. That is, larger animals were hotter than would be expected by surface/volume considerations.

Is there a more subtle mechanism that can account for this difference? Consider the process by which an organism acquires the resources used to fuel its metabolism. Usually these resources arrive through one or more openings (mouth, nose, or the like) and must be distributed throughout the organism's interior. The transport problem would be overwhelming if there had to be

16

a direct path from the opening(s) to every cell. Instead, the resources are distributed by a branching network. What, then, is the form of a branching network for efficiently distributing the resources? Consider the idea of moving a fractal surface to the interior of the organism; that is, consider a transportation network that branches in a self-similar way. Such a network will distribute resources with much less 'tubing' than one which has a direct path to every cell. At the end of the 20th century, J. H. Brown, B. J. Enquist, and G. B. West (as discussed in Mitchell's book *Complexity*) carefully developed this fractal approach and, lo and behold, the 3/4 power law emerged. In short, the metabolic rate was limited by the efficiency with which the organism could distribute resources to the cells.

The occurrence of a power law is often taken as an indicator of complexity, but caution should be exercised: it is neither a necessary nor a sufficient condition for complexity. Nonetheless, the kinds of scaling just examined, and the related networks for distributing critical quantities, are closely coupled in the study of complex systems.

Networks

Networks, in particular, have become an important tool for studying complexity. Network theory, more formally known as *graph theory*, was put on a firm basis by Denes Konig in the early 20th century, though the use of particular networks, such as inheritance *trees*, goes beyond Darwin to early times. Several collaborations centring on Duncan Watts and Mark Newman (see Mitchell's *Complexity*) advanced the study of networks to the point of characterizing properties of networks observed in complex systems. Two commonly occurring networks are characterized as *scale-free* networks and *small-world* networks. The just-described 'fractal' resource distribution network is an example of a *scale-free* network. A *small-world* network is arranged so that (1) most nodes in the network are connected only to nearest neighbours, but (2) there are a few carefully selected long-range connections between clusters of nodes (see Figure 3). The oft-cited '6 degrees of separation' between any two people in the world (say you and the president of the USA) is a typical property of small-world networks. The advantages and limitations of network theory will be discussed more thoroughly in Chapter 4.

Distances between nodes:

Net 1, each node:
 max. distance = 8
 ave. distance = 36/8 = 4.5

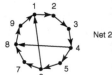

Net 2, each node:
 max. distance = 7
 ave. distance = 45/9 = 5

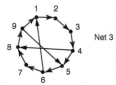

Net 3, node 1:
 max. distance between node 1 and
  other nodes in the network = 6
 ave. distance between node 1 and
  other nodes in the network = 3.75

Net 3, node 2:
 max. distance between node 1 and
  other nodes in the network = 5
 ave. distance between node 1 and
  other nodes in the network = 3.375

Net 4:
 max. distance between nodes = 6
 ave. distance between nodes = 3.75

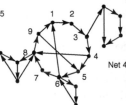

3. Small-world networks

Dynamics

As mentioned in Chapter 1, the physicist's concept of *state* provides the foundation for modelling the dynamics of CPS.

The state $S(t)$ of a system at time t summarizes the results of activity in the system prior to time t; usually this is done in a way that makes it possible to predict future possibilities knowing just the information provided by the state. The game of chess provides a useful analogy. It is sufficient to know the present configuration of pieces on the board to know what moves are possible from that time onward under the rules of chess; it is not necessary to know the history of how that configuration was obtained. Similarly, the state of a gas, typically its <pressure, temperature, volume>, provides sufficient information to determine its dynamic changes from that time onward under the gas laws. Newton used the current positions of masses in a system to predict future positions of the masses under the laws of gravity. The resulting theory was so general that it is used currently to determine the complicated trajectory of a planetary probe, a use that could not possibly have been anticipated when Newton formulated his theory. Maxwell used a similar (more subtle) notion of state to build a theory of electromagnetism, again with unanticipated applications such as television, computers, and the Internet.

Almost all CPS studies approach dynamics with a careful definition of state, but emergence complicates the analysis. The Ising problem (see Figure 4) provides a simply defined example of the complications introduced by emergence. Consider a lattice (e.g. a checkerboard) with one bar magnet ('north pole'/'south pole' at opposite ends) occupying each position of the lattice. The definition of state in this case is simple: it is the orientation of each magnet at each lattice point. The question is, 'What orientations will the magnets eventually take if they start out with poles randomly oriented?' One possibility would be for all the component magnets to line up with the north pole of any given

Bar magnet array at stable equilibrium:

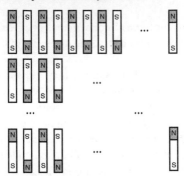

Bar magnet array with persistent broken symmetry ('frozen glass')

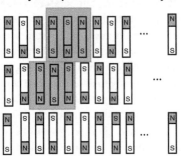

4. **Ising model**

magnet opposite the south pole of its adjacent neighbours (taking account of the fact that the north pole of one magnet attracts the south pole of another). However, in practice, the magnets often line up in small domains where, at the edge of one domain, north poles are opposite north poles in the adjacent domain. That is, instead of a system with a uniform overall symmetry, we have a broken symmetry provided by many bounded domains of different orientation. The explanation of the emergence of these broken symmetries requires a careful, difficult mathematical analysis (too intricate to be recapitulated here). Roughly, as the bar

20

magnets attempt to align, they 'freeze' into a glass-like state, much as a hot liquid glass becomes a solid with a fluid-like organization (liquid water is amorphous, while frozen water is crystalline).

Symmetry-breaking also plays an important role in the development of multi-celled organisms and, indeed, in complex systems in general. Alan Turing, the great mathematician and computer pioneer, in one of his final papers, investigated the complicated processes whereby a fertilized egg develops into a multi-celled organism with many different kinds of specialized cells, a process called *morphogenesis*. What mechanisms enable an egg cell, which reproduces by fission, to produce cells that are not copies of itself? In physicists' terms, how does the egg cell break the symmetry of the copying process?

Symmetry-breaking in living organisms is familiar at many levels. The cubs of both leopards and lions are spotted, but the mature lion has no spots. How does the spotted coat form to begin with, and why does it later disappear in one case and not the other? At a much lower level, the behaviour of ordinary slime mould under wet and dry conditions exhibits a surprising instance of symmetry-breaking. Under wet conditions, slime mould cells move about individually in amoeba-like fashion, searching for food. However, if the region in which they are living begins to dry up, the individual cells aggregate to form a base, and then go on to form a stalk capped by a capsule full of spores. All of the cells involved in the base, stalk, and capsule then die except for the spores. The spores are distributed throughout the area, and can subsist for years until wet conditions prevail again, at which point each spore turns into the free-living form again.

The symmetry-breaking in slime mould is at once simple to describe observationally and complex to explain in terms of molecular mechanisms. It involves both a 'slow' and a 'fast' dynamic. The 'slow dynamic' is a quintessential evolutionary adaptive process that provides the building blocks for morphogenesis. It will be discussed in the next chapter. The 'fast dynamic' is nicely exemplified by Turing's model. Turing used differential equations well-known to physicists to specify a combination of laws for reactions (classical chemistry) and laws for diffusion of the reactants. There are just two reactants, C (catalyst) and I (inhibitor). C catalyses the formation of both itself and I; I inhibits the formation of C and also diffuses more rapidly than C. A simulation shows that, starting from a homogeneous system, this reaction/diffusion dynamic yields local peaks of C surrounded by I (see Figure 5). The differential rates of flow of C and I play a key role in breaking the initial symmetry.

Many CPS problems (e.g. the flow of electrons in superconductive materials) also involve flows—flows that are nicely described by networks. Networks provide a detailed snapshot of CPS and complex adaptive systems (CAS) interactions at any given point in their development, but there are few studies of the evolution of networks (the slow dynamic). Indeed most of the tools discussed to this point concentrate on the *results* of adaptation, rather than on the *process* of adaptation. The distinction between the fast

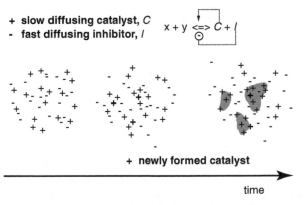

5. Turing's model of symmetry-breaking in morphogenesis

dynamic of flows (change of state) and the slow dynamic of adaptation (change of the network of interactions) often distinguishes CPS studies from CAS studies. The next chapter will examine tools tuned to CAS, where adaptation has a critical role.

Chapter 3
Complex adaptive systems (CAS)

Characteristics of complex adaptive systems

Complex adaptive systems (CAS) are composed of elements, called *agents*, that learn or adapt in response to interactions with other agents. Markets, as mentioned in Chapter 1, provide a familiar example. From the time of Adam Smith onward, markets have been the object of careful study, with Kenneth Arrow and Gerard Debreu setting forth a major mathematical theory (a theory for which they won the Nobel prize). That theory demonstrates that a market made up of 'fully rational' traders will 'clear'—that is, the market will attain a stable price based on supply and demand, with only small fluctuations that provide no consistent way for traders to make profits. In CPS terminology, the market attains a *steady state*.

However, the Arrow-Debreu theory does not take into account adaptive interactions typical of a CAS. From the CAS viewpoint, the 'fully rational' agent assumption is a *very* strong assumption. Each agent must act on full knowledge of the *future* consequences of its actions, including the responses of other agents to those actions. Clearly no realistic agent possesses such omnipotence. Arrow was aware of this difficulty from the start, pointing out that real markets involve diverse traders of *bounded* rationality, with different agents employing different strategies. Moreover, realistic agents change their strategies as they gain experience with the diverse actions of other traders—they adapt. Markets made up of such agents rarely reach an equilibrium, even temporarily; rather, there are often large fluctuations ('bubbles' and 'crashes') caused by the traders' ongoing, diverse adaptations.

The *diversity* that results from continuing adaptation is a hallmark of CAS, and it presents a difficulty not easily overcome. The extent of this diversity was pointed up in the rainforest example at the start of the first chapter, and it is equally well exemplified by diversity of proteins in a biological cell (examples we'll examine more closely in Chapter 5). The difficulty presented by diversity is further compounded by the conditional nature of agent actions—the actions of a typical agent are conditionally dependent upon what other agents are doing. Said another way, the behaviour of the whole CAS is not obtained by *summing* the behaviours of the component agents or, using a familiar phrase, 'the whole is more than the sum of the parts'. This non-linearity (non-additivity) poses particular difficulties for the mathematics used to study CPS because most of that mathematics (e.g., partial differential equations) is founded on an assumption of additivity, To add another difficulty still, the aggregate behaviour of a CAS—think of swarms of fish or flocks of birds—is rarely set by a central executive.

Despite these difficulties, a theory of CAS is possible. The possibility stems from features that are common to all CAS that have been carefully observed. For instance, all well-studied CAS exhibit *lever points*, points where a small directed action causes large predictable changes in aggregate behaviour, as when a vaccine produces long-term changes in an immune system. At present, lever points are almost always located by trial and error. However, by extracting mechanisms common to different lever points, a relevant CAS theory would provide a principled way of locating and testing lever points. Other known mechanisms and common features extend the opportunities for theory; they will be the concern of this and following chapters.

Agent structure

All CAS agents, whatever their particularities, have three levels of activity:

1. Performance (moment-by-moment capabilities)
2. Credit-assignment (rating the usefulness of available capabilities)
3. Rule-discovery (generating new capabilities).

Performance designates an agent's behavioural repertoire at a point in time, typically specified by a set of conditional IF/THEN rules. In a simple case, a rule might be:

IF (there's an approaching object in the visual field) THEN (flee).

In more sophisticated cases, a rule will send signals that can activate other rules in the performance set. For example, in a biological cell, one gene in a chromosome may cause the formation of a signal protein that turns on other genes in the chromosome. More generally, an agent retains a list of signals currently being sent (such as the signal proteins contained in a biological cell), and rules would be of the form:

IF (signals x and y are present on the list) THEN (add signal w to the list).

Signals sent and received by rules can be used to chain rules together to form sequences and subroutines—much as is done in computer programming. It is not difficult to select a small set of elementary IF/THEN rules, like the machine-level rules of a general-purpose computer, that can be chained together to implement any possible programmable performance. This generality makes it possible to apply the signal-processing approach to any CAS composed of agents that can be described via computer simulation.

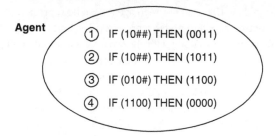

Agent

① IF (10##) THEN (0011)

② IF (10##) THEN (1011)

③ IF (010#) THEN (1100)

④ IF (1100) THEN (0000)

**Corresponding signal-processing
network**

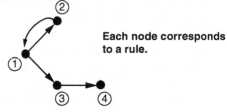

**Each node corresponds
to a rule.**

6. An agent and the corresponding network

To make an agent sensitive to its environment, it is provided
with a set of *detectors* (sight, smell, and touch for example) that
translate activities in the environment (say, the activities
of other agents) into signals for internal processing. The agent
can also modify the environment via a set of *effectors* (think
of muscles) that translate the agent's interior signals into
actions that affect the environment (see Figure 6). With these
provisions, signals in different applications can represent
anything from cash in markets or resources in an ecosystem
to impulses in a nervous system or digital packets on the
Internet. Indeed, one of the advantages of the signal-processing
approach is that signals of different kinds can interact, as
would be required, say, when the CAS concerns the relation
of markets to sustainability.

The flow of signals within and between agents can be represented by a directed network, where nodes represent rules, and there is a connection from node x to node y if rule x sends a signal satisfying a condition of rule y. Then, the flow of signals over this network spells out the performance of the agent at a point in time. Thus, for example, the flow of spikes (impulses) over the neural network of a mammal determines its behaviour. The networks associated with CAS are typically highly tangled, with many loops providing feedback and recirculation, a point that will be discussed at greater length in the next chapter. An agent adapts by changing its signal-processing rules, with corresponding changes in the structure of the associated network. These adaptive changes are the concern of the next two topics.

Credit assignment. As an agent accumulates experience, it will find that some of its rules are rarely, or never, useful. If a frog flees from all moving objects it will rarely eat or mate. In order to adapt, an agent requires a means of assigning a quantity, usually called *strength*, that rates the usefulness of different rules in helping the agent to attain important resources, such as food, shelter, and the like. It is easy enough to assign high strength to a rule when that rule's action immediately gives rise to a desired result—the classical reinforcement of Pavlovian conditioning. But how is credit assigned when an action merely 'sets the stage' for subsequent useful actions? It is especially difficult to assign credit to a stage-setting rule when the action it specifies entails a cost, as in the sacrifice of a piece in chess to make possible a powerful later move.

One approach to assigning credit to stage-setting rules is to set up a market where rules buy and sell signals, using their strengths as 'cash in hand' for this purpose. Consider then a chain of rules, where each rule is activated by a signal from its precursor in the chain. In this market, each rule in the chain must pay its predecessor, from its cash in hand, for the right to send its signal. It then receives a payment in turn from its successor. Thus a chain of rules acts much like a sequence of 'middlemen' leading from an initial situation (think of a mining operation) to a final situation (think of an automobile buyer). A rule grows stronger—attains more cash in hand—if it makes a profit by selling its signal for more than it paid for the incoming signal.

New cash enters a chain only from a 'reward' acquired by the final rule in the chain when it causes the acquisition of a desired resource (think of the sale of an assembled automobile). That is, as in Pavlovian conditioning, the strength of the final rule is increased when it produces an immediate reward. Now, set up the market so that each rule pays a fixed proportion of its strength to its predecessor. Then, on the next activation of the chain, the final rule will pay more to its predecessor because of its increased strength. Subsequent activations will cause this pulse of increased profit to move back up the chain, as in real chains of middlemen. After repeated activations, the earliest rules in the chain will share in the rewards attained by the final rule(s) in the chain. The process, of course, can also cause losses in strength: if the final rule loses strength, all rules in the chain become weaker.

This marketing approach, called a *bucket brigade* algorithm in the literature, has the great advantage that each rule need only be 'aware' of the strengths of its immediate predecessors and successors (Figure 7). There is no need for the computationally expensive task of tracing all possible paths from early-acting rules to later-acting rules. Instead, the rule's strength is determined by its direct interactions with (1) the rules sending signals to it, and (2) the rules directly acting on its signals. A rule becomes stronger when it is a middleman in one or more chains that repeatedly lead to a valuable end-product, even though the rule's 'position' in the chain is never calculated.

Rule discovery. When an agent treats rules as hypotheses that must be tested, then credit assignment amounts to a process of confirmation: strong rules are hypotheses that have been (partially) confirmed, while weak rules are hypotheses that

A chain of rules

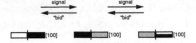

A rule's bid is 20% of its strength.
A rule active at the time a reservoir is filled receives a payment of 100.

t−1	80	100	100
t	100	80	100
t+1	100	100	80
t+2	100	100	180
Next repetition			
T	100	80	180
T+1	100	116	144
T+2	100	116	244
2nd repetition			
T	103	93	244
T+1	103	142	195

7. Successive changes of strength under a *bucket brigade*

have been disconfirmed. If an agent can replace weak rules with new rules that might *plausibly* be better, it can adapt at a much higher level. But how are *plausible* new hypotheses generated?

Because rules work together much like subroutines in a computer, using a randomly generated rule to replace a weak rule is about as likely to be useful as randomly replacing an instruction in a computer program. The chances of an improvement are vanishingly small. Somehow the replacement must be biased by the agent's previous experience.

Building blocks come into play at this point. It has been observed that innovation in CAS is mostly a matter of combining well-known components in new ways. For example, the internal combustion engine, a major influence on 20th-century economies, was a new combination of well-established building blocks: gears, wheels, pumps, Volta's sparking device (sparkplugs), Bernoulli's perfume sprayer (the carburettor), and so on. Maxwell's elegant theory of electromagnetism was based on metaphors using tow-lines (lines of force), wave propagation in canals (fields), etc., a fact recorded in his collected papers. This use of well-established building blocks incorporates experience in a way that substantially increases the plausibility of a new hypothesis or rule.

In extracting building blocks for signal-processing rules, it is helpful that they can be represented as strings over a fixed alphabet, much like specifying chromosomes as strings over four nucleotides. (Representation by strings will be discussed at greater length in Chapter 6). The strings representing different rules can then be compared to locate common substrings, again as in molecular genetics, where common substrings are called *motifs* (recalling motifs in musical composition). Motifs common to several strong rules amount to building blocks that have had their usefulness established in several contexts.

New combinations of motifs can be generated by 'cross-over' rules, in a way similar to the way a breeder obtains new combinations of desirable characteristics by cross-breeding animals (Figure 8). A horse-breeder, for example, crosses a horse with long legs with a horse that has a strong neck to get (some) offspring that have both long legs and a strong neck. Similarly, strong rules can be cross-bred to produce a pair of 'offspring' rules that have new combinations of the motifs they contain (see Chapter 6 for more detail). The new rules so produced are then tested under credit-assignment. If an offspring rule attains sufficient strength it will in turn be cross-bred. Over successive generations, with strength serving as Darwinian fitness, weak rules will typically be replaced by progressively stronger rules that have new combinations of building blocks common to strong rules.

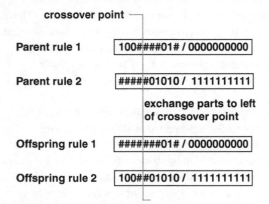

Rule format

condition   / action (signal)

| 100####01# / 00000000000 |

crossover point

Parent rule 1 | 100####01# / 0000000000 |

Parent rule 2 | ####01010 / 1111111111 |

exchange parts to left of crossover point

Offspring rule 1 | ######01# / 0000000000 |

Offspring rule 2 | 100##01010 / 1111111111 |

**Offspring 1 is more general (has more #s) than either parent.
Offspring 2 is more specific than either parent.**

8. Producing new rules by cross-over

CAS theory

The study of CAS is still in its earliest stages, so only pieces of a theory exist. Still the pieces that exist do suggest the possibility of an overarching theory. One characteristic common to all CAS points the way: the behaviour of a CAS is always *generated* by the adaptive interactions of its components. And the hierarchical structure characteristic of CAS is also generated—particular combinations of agents at one level become agents at the next level up. The organization of a biological organism provides a familiar example:

1. The chromosomes, through a translation apparatus, generate the proteins that serve as low-level agents (catalysts, signals, etc.).

32

2. The proteins combine to form membrane-enclosed organelles that act as agents for higher level processing.
3. Organelles combine to form cells.
4. And so it goes, through organs, organisms, co-evolving populations, etc.

Such hierarchical generative processes characterize all CAS.

Generative processes suggest formal grammars (see Chapter 1) as a starting point for an overarching theory. Recall that a formal grammar involves two basic parts: an *alphabet* (letters or elements) and a set of *operators* (that are applied to letters and strings of letters to produce new strings). That *bête noire* of elementary school, sentence diagramming (parsing), was a (mostly informal) way of defining legitimate sentences using grammatical notions. Euclidean geometry provides another familiar, more formal, example, where the axioms provide the starting 'alphabet' and the rules of deduction constitute the 'operators' (generating new theorems from the axioms and theorems already produced). Other examples are the rules for combining atoms based on the periodic table of the elements, the use of a group operation (say multiplication) to define the elements of a mathematical group (say the natural numbers), and aforementioned physical mechanisms (e.g., the fluid flow equations) that underlie changes in state (e.g., pressure, temperature, and volume for a gas). Chapter 6 will further discuss the relation between grammars and mechanisms in the CPS/CAS context.

The corpus generated by a grammar (e.g., all legitimate English sentences, all theorems of Euclidean geometry, all inorganic chemical compounds, and so on), and the properties associated with the elements of the corpus, *emerge* from the repeated action of the grammar's operators. There is an intrinsic time-like order in which the objects in the corpus emerge, starting with the alphabet and listing the objects that can be formed by one application of the operators, two applications of the operators,

33

and so on. For dynamic systems, as mentioned earlier, we can think of this order of generation as successive (possible) time-like changes in state.

For CAS analysis the formal definitions provided by a grammar are a good starting point, but formal definitions do not a theory make. It remains to supply enough analysis to allow prediction, or at least suggest where to look for mechanisms that underpin critical features (such as the lever points mentioned earlier in the chapter). The difficulty of analysing grammar-defined systems is pointed up by the fact that, after more than 50 years of study, we still have few tools for predicting the performance of computer programs, though the grammars in that case are typically quite simple.

Still, there are possibilities. One of these possibilities stems from one of the difficulties: the hierarchical organization of CAS. This hierarchy is determined by a succession of enclosing boundaries that pass some signals and not others, e.g. the semi-permeable membrane hierarchies of biological cells. The content of each bounded enclosure consists of the resident signals (e.g. the membrane-contained proteins in the case of a biological cell). For many CAS, the interactions within an enclosure can be approximated by a 'billiard ball' chemistry, where interactions are determined by random collisions between the signals and rules therein (see Figure 9). Think of the different rules and signals as being identified with different colours, so that the processing of a signal causes a change in the colour of a signal ball when it collides with a rule ball. Under this regime, the rate of processing of a given kind of signal depends upon the proportions of the relevant rules and signals within the enclosure, as in elementary chemistry.

This billiard ball model lets us bring into play the 'urns' of elementary probability theory, where the probability of randomly drawing, say, a black ball from an urn is set by the proportion of

Two irreversible reactions separated by a semi-permeable membrane that only passes the product of the first reaction:

$$A + B \Rightarrow Y \qquad Y \underset{\shortparallel}{\overset{\shortparallel}{\Rightarrow}} \qquad Y + Y \Rightarrow C$$

A and B are replenished as soon as used; Y is removed as soon as formed.

C is removed as soon as formed.

○ A
● B
◐ Y
● C

Urn 1 — gate lets Y exit ◐

Urn 2 — gate lets Y enter — gate lets C exit ●

Concentrations of A, B, and Y are kept equal in urn 1

$$p(A) = p(B) = p(Y) \, 1/3$$

Urn 2 contains only Y

$$p(Y) = 1$$

9. Urn model of coupled reactions

black balls in the urn. To simulate a random collision of two balls we draw a pair of balls at random from the urn. Looked at in this way, a collection of urns can represent a collection of boundaries, each of which encloses a set of catalysts and signals. To simulate the diffusion (movement) of balls between urns, we draw a ball at random from one urn and place it in another. We can, in addition, constrain the diffusion by supplying each urn with entry and exit 'gates', so that only designated colours are passed in and out of that urn—a random draw of ball that cannot pass is simply placed back in the urn. With this provision each urn acts like a semi-permeable membrane, passing some signals and not others.

By putting signal/boundary hierarchies into this gated urn format, we can bring to bear an important part of probability theory, the theory of Markov processes. Briefly, in a Markov process, one state changes into another in a probabilistic fashion, rather than always changing into a fixed following state. Random collisions between billiard balls, as in the billiard ball version of chemical reactions mentioned earlier, are well-described by a Markov process.

The theory of Markov processes is well-developed (more about this in Chapter 6) and it offers powerful, relevant tools for examining gated urn models, and through them CAS. In particular, Markov theory can provide an explicit description of the likely distribution of signals under a specified performance regime. That, in turn, provides insight into the possibilities for further adaptation. Though this use of Markov processes is far from an overarching theory of CAS, it does suggest a useful path.

Chapter 4

Agents, networks, degree, and recirculation

Agents and networks

When studying complex adaptive systems (CAS) in a grammar-like way, agents serve as the 'alphabet'. In agent-based terms, then, the hierarchical organization of CAS implies different kinds of agents at different levels, with correspondingly different grammars. At the lowest level, there are individual signal-processing rules generated as strings over an alphabet having as few as three or four symbols (as discussed in the next paragraph). At the next level, there are agents specified by a collection of rules, where the corpus generated by the grammar consists of sets of finite rule collections. At still higher levels there are groups of interacting rule-based agents, as in a multi-celled organism or a market.

It's helpful at this point to be more explicit about signal-processing rules. As discussed earlier (in Chapter 3), signal-processing rules have a condition/action form: IF (signal x present) THEN (send signal z). Signals can generally be encoded as a binary string of 1s and 0s. It is also true, though less familiar, that the IF condition for accepting binary strings can be specified using one additional symbol #, sometimes called a 'don't care' symbol. In the simplest use of this symbol, the string 1## specifies an IF condition that will accept any three-digit binary string that starts with a 1, while ###01 accepts any five-digit binary string that ends in 01. In this format, IF(1##)THEN (00001) sends the signal 00001 whenever a string starting with a 1 is present; the rule IF(###01)/THEN(01111) would then process the signal 00001 to produce the signal 01111, and so on. There are simple extensions of this {1,0,#} alphabet, and the rule format, that increase flexibility—for example, the rule may be of the form IF(signal x present AND signal y present) THEN (send signal z), requiring signals x and y to be simultaneously present before the output signal z is produced. But a simple alphabet and grammar of this kind can be used to generate a corpus of rules that, acting together, can implement any programmable signal-processing agent. In computer science terms, the corpus generated is *computationally complete*.

As described under *performance* in the previous chapter, the interactions of signal-processing agents at a point in time can be specified by a network—a snapshot of the agents' performance capability. The richness of an agent's interactions is given by the number of outgoing *edges* (connections) from the *vertex* (node) representing the agent, a network property called *fanout* or *degree*. A typical machine (say a computer) has a fanout of around 10, while many CAS (e.g. the human central nervous system) have a fanout of 1,000 or more. Because we have yet to build a machine with large fanout, there is little basis for speculating about the capabilities of such machines (e.g. it is moot whether or not such a machine could be 'conscious').

The study of networks in terms of fanout leads to the definition of *communities* within the network. A *community* is a set of nodes where the connections of each node in the set largely lead to other nodes within that set. The quantification of 'largely' determines the 'tightness' of the community. A large set of nodes with 50 per cent of the connections leading to other members of the set is a quite different community from a small set of nodes with 90 per cent of the connections leading back into the set. Indeed, a loose community may contain several tight communities, leading on to community-oriented definition of hierarchical organizations.

Issues of community play a key role in practical Internet activities, such as income from advertising.

Loops

The combination of high fanout and hierarchical organization results in complex networks that include large numbers of sequences that form loops. Loops recirculate signals and resources (as when trees absorb water from the ground, which then evaporates from the leaves and ultimately returns to the ground through rainfall). Loops also offer control through positive and negative feedback (as in a thermostat). Equally important, loops make possible program-like 'subroutines' that are partially autonomous in the sense that their activity is only modulated by surrounding activity rather than being completely controlled by it. Such autonomy allows the CAS to run internal activities that go beyond the current stimulus. More complex loops allow the CAS to 'look ahead', examining the effects of various action sequences without actually executing the actions (as in examining alternatives lines of play in a game).

The complexities introduced by loops have so far resisted most attempts at analysis. Over a half century ago Donald Hebb outlined a sophisticated theory of human behaviour based on tangled loops in neural networks called *cell assemblies*. About the same time Arthur Samuel introduced a sophisticated way of combining 'lookahead' and credit assignment in a checkers-playing computer program (a program that learned to beat Samuel himself). Though there are large amounts of data in both cases, we still lack analytic theories of these models. In both cases, lookahead plays a key role, and in both cases this capacity is steadily modified by learning. From the network point of view, this learning produces a series of networks (snapshots) indicating the changes produced. The ontogeny of these changes—understanding how they originate and develop over time—is indispensable to the analysis of lookahead and related loop-based behaviour. Subsequent chapters will look more closely at ontogeny in relation to adaptation.

Tree-like approximations and evolutionary games

The difficulties of analysing the behaviour of networks with many interior loops has, both historically and currently, encouraged the study of networks without loops called *trees*. Trees occur naturally in the study of games. As outlined in the last part of Chapter 1, each configuration of pieces on the board is represented by a node, and each legal move from that configuration is represented by a connection to the node representing the resulting changed configuration. Starting from the initial configuration, the result is a directed graph without loops, each path through the graph terminating in one of the game's final configurations. The average fanout per node is typically related to the difficulty of the game.

More generally, as with a computer subroutine (before it loops back on itself), or the 'feedforward' parts of a neural network, one can extract parts of generated structures that are trees. Indeed, because trees are easier to analyse, most artificial neural networks constructed for pattern recognition are trees. The detectors produce pulses representing properties of the input scene; these pulses propagate through successive levels of the tree, ultimately activating output effectors that identify patterns in the input scene. If the tree has k levels, then the pulses will reach the effectors in k steps. Because the pulses pass directly from level to level (the feedforward provision), pulses from a scene that is presented k steps later cannot interact with pulses from the first scene. This greatly simplifies analysis, but it forgoes direct comparison of successive patterns—the 'working memory' provided by circulating pulses. This loss is analogous to modelling an ecosystem without allowing recirculation of resources.

Evolutionary game theory makes use of the tree structure of games to study the ways in which agents can modify their strategies as they interact with other agents playing the same game. Samuel's technique, using credit assignment to modify a checkers-playing agent's strategy, offers a striking example.

However, evolutionary game theory does *not* concern itself with the evolution of the game's laws. For example, the appearance of new species in an evolving ecosystem offers new opportunities for interaction, but such possibilities fall outside the evolutionary game theory framework. More generally, evolutionary game theory cannot be used to study a basic feature of CAS: the origin and exploitation of hierarchies. The origin of hierarchies will be the concern of the next two chapters.

Chapter 5
Specialization and diversity

Specialists

A multi-celled organism, anything from a small nematode to a large primate, consists of specialized cells organized into a variety of communities. In a primate, for example, there are organs like the heart, lungs, various muscle sets, brain, and so on, each specialized to carry out a distinct activity such as pumping blood, adding oxygen to the blood, manipulating the environment, etc. Although counterparts of all these activities can be found in a single free-living cell, such as an amoeba, communities of specialized cells allow more efficient execution of the activities (as we'll see shortly).

Community-based organization extends both downward and upward from the level of organisms, giving rise to the hierarchies that characterize complex adaptive systems (CAS). For example, an individual biological cell is delimited by an outer membrane; within that membrane are further membranes delimiting organelles, the nucleus, etc. The nucleus in turn contains clusters of proteins centred on DNA. All this apparatus greatly enhances the manipulation of the cell's basic building blocks, proteins. In the upward direction, there are ecosystems where the term *niche* is used to describe communities of interacting species (more about this in Chapter 7), and on up to continent-wide clusters of ecosystems.

When one looks at biological communities, it is amazing just how specific some interactions can be. Darwin's 'comet orchid' provides an example that also shows that the theory of natural selection *can* be used to make strong predictions. Darwin made a careful study of orchids and knew that orchids related to the comet orchid are pollinated only by moths. Because the nectar tube of the comet orchid is a foot long (hence the name), Darwin surmised that there must be a moth with a foot-long proboscis (!) that specializes in taking nectar from (and pollinating) the comet orchid.
The moth involved, a variety of hawk moth, was finally filmed over a century after Darwin had made this prediction. Because of their extreme co-evolutionary adaptation, neither the moth nor the orchid could survive without the other. Counterbalancing this weakness is the fact that the moth has no competitors for its food source, and the orchid wastes no nectar on inefficient pollinators.

Communities provide for the interaction of specialists, but what advantage does this interaction have over the independent activity of an equivalent number of generalists? Adam Smith's famous example provides an important clue. Just prior to the time Adam Smith was writing, straight pins were produced by blacksmiths (generalist craftsmen) through a combination of skills: drawing wire from molten metal, clipping the wire to produce a sharp point, and adding a blunt head to the other end of the wire.
The process was difficult and time-consuming, making straight pins a luxury item. Then, at the time Adam Smith was writing, one of the first 'production lines' formed for the purpose of producing straight pins, each step in the process being carried out by a specialist. Throughput increased by a factor of 10, and straight pins became widely available.

Similar increases in throughput are observed when one looks at membrane-separated cascades of interactions within a biological cell. The semi-permeable membranes enclosing organelles admit some proteins and not others. The immediate effect is an increase in the concentration of the admitted proteins within the organelle.

By the laws of elementary chemistry, the higher concentrations increase the reactions between the admitted proteins. Other organelles in the cascade can then process the products of these reactions, yielding a production-line-like cascade of specialists, à la Adam Smith. The result is a substantial increase in the availability of products involved in the cell's survival and replication, thereby increasing the cell's Darwinian fitness.

Diversity

All CAS that have been examined closely exhibit trends toward increasing numbers of specialists. As an example, consider the progression of changes from the market squares of early towns to present-day commodity markets. In the early markets, individuals exchanged self-produced products, such as hand-woven cloth, eggs, leather, and so on. A present-day commodity market consists of a vast array of specialists dealing in futures, hedges, derivatives, etc., none of which involves handling the actual commodity. Looking to other CAS, the rampant specializations in contemporary automotive production lines or government bureaus are a far cry from their earliest versions. Similar observations apply to the Internet, weather bureaus, and flight control, to name a few others.

In every case, there are diverse specialist agents that attend to, and process, selected signals. As in biological cells, the diversity is sustained by a combination of boundaries (the cell's semi-permeable membranes), signals (the proteins), and signal-processing (the reactions between proteins). A surprising characteristic of this signal-processing is that relatively small parts of signals, which I will call *tags*, route the signals through the boundaries. Tags have different names in different CAS studies—'active sites' (proteins), message headers (the Internet), 'motifs' (molecular genetics), and grammatical declensions (languages), and so on—but the 'routing' functions are similar. For example, small parts of protein signals cause them to adhere to specific parts of chromosomal DNA, turning genes 'on' and 'off' and enabling complex signal-dependent computations. This tag-based control allows cells to exist in a diverse range of conditions by activating only genes that are relevant to the current situation. Similar advantageous relations between tags and boundaries can be found in CAS of all kinds.

Tags have a particularly important role when they are used to coordinate the rule sequences exemplified by cascades of reactions, 'production lines', and the 'middleman' sequences discussed under *credit assignment* in Chapter 3. When tags coordinate 'production lines', variations with substantial differences in throughput and efficiency are easily generated and tested. Because many kinds of production-line-like cascades are possible, CAS agents can develop ever more complex strategies for survival. And each new kind of agent offers still further opportunities for interaction. In this increasing set of possibilities for interaction we have the beginnings of an explanation for the pervasive diversity of CAS—new specializations open the way for still further specializations and still greater throughput and diversity.

To get a more precise view of the routing possibilities for tags, let's look again at the use of #s to define the conditions for signal-processing rules (introduced at the beginning of Chapter 4). A condition can be set to respond to a given tag by using #s (don't cares) on either side of the tag; thus the condition #100###...# responds to a signal string with tag 100 at the 2nd, 3rd, and 4th positions of the string. Note that a condition with many #s can accept a variety of tags while a condition with few #s will accept few if any tags (requiring a specific string when it has no #s). Or, looking at tags themselves, short tags satisfy a variety of conditions, while long tags make highly specific requirements on conditions.

In a binary coding, only a few positions are required to provide for hundreds of distinct tags (ten positions suffice for over 1,000 distinct tags). Thus, tag diversity is easily achieved, and relevant routings are easily maintained. Moreover, interactions can be simply re-directed by making modest changes to tags. The simple relation between generality and specificity under the # notation offers a direct way of transforming rules, and the agents containing them, from generalists to specialists, or vice-versa. The question then becomes: How does a CAS discover or generate relevant tags and the sequential routing that yields 'production lines'?

Origins of production lines

As just pointed out, it is simple to modify tags, and the conditions that route the tags through boundaries, to generate a wide range of interactions. The problem is to do this in ways that build on an agent's experience. Cross-breeding, discussed earlier in connection with rules (under 'Rule discovery' in Chapter 3), offers a good way to accomplish experience-based generation of new tags. Consider two rules that are crossed in the tag-region of their condition (signal-receiving) part. Let the two rules each accept a distinct tag, and let the tags occupy overlapping loci in the string, with one condition requiring a more specific tag (more non-#s) than the other. The result of a cross within the region of overlap between the tags will be two new rules, with condition parts requiring new tags (as a result of the exchange of parts). Generally, one of the condition parts accepts a wider range of tags than either parent (it has more #s in the tag region), while the other is more specific (see Figure 8).

A rule-condition with a more general tag requirement can accept signals in addition to those accepted by either parent. In particular, this new rule may accept signals from another production line. The new rule becomes a new buyer for some rule in the other production line, thereby supplementing that production line's income. If the new rule sets the stage for a new use of the original production line, and the resulting throughput is valuable, then both the production line and the new rule will prosper. Under the bucket brigade credit-assignment procedure

('Credit assignment' in Chapter 3) there will be corresponding
increases in the strength of all the participating rules, thereby
assuring the survival of the augmented system.

The same procedure can generate new production lines 'from
scratch'. Consider a rule that often gets immediate rewards but
has no coupled supplier, so that it is often 'unprepared' when
reward opportunities occur. Then, let cross-over generate
a rule with a signal (resource) that satisfies the condition of the
rewarded rule. The newly generated rule thus 'sets the stage' for
activation of the reward-acquiring rule, making it more likely to
be rewarded when activated. When this happens, the coupled
pair acts as a two-agent production line. The two-agent line
then becomes a candidate for further stage-setting, providing
opportunities for still longer lines.

Generally, when the output of a production line is valuable it is
used up rapidly—as when an agent sequesters the output to help
it reproduce. If we look upon the production line as a sequence
of chemical-like reactions, then 'valuable' end-products of the
sequence are rapidly sequestered, lowering their ambient
concentration. The resulting lower concentration of end-products
lowers the back-reaction rate (disassembly)—much as in Herb
Simon's classic comparison of two watchmakers. Thus, under the
laws of elementary chemistry, there is an increase in net
(forward) throughput of reactions leading to valuable,
sequestered end-products. In other words, production lines with
valuable outputs are more efficient at acquiring and processing
relevant reactants, so they tend to dominate an agent's activities.

In CAS agents, many rules and production lines are active
simultaneously (recall the activities in a biological cell). Because
new rules generated by cross-over do not usually replace their
parents (replacing weak rules instead), extant production lines
are rarely disrupted by new rules, allowing exploration without
disrupting regularities already being exploited. Under this

procedure for generating new rules, tags serve as building blocks that can be recombined to foster new production lines. The next chapter will take a closer look at the relations between building blocks, generated systems, and the phenomenon of emergence.

Chapter 6
Emergence

Building blocks revisited

As discussed earlier in this book, the characteristic of 'wetness' cannot reasonably be assigned to individual H_2O molecules, so we see that the wetness of water is not obtained by summing up the wetness of the constituent molecules—wetness *emerges* from the interactions between the molecules. Similarly, common properties of markets, such as 'bubbles' and 'crashes', are not well-described by summing (say averaging) the acts of individual traders—these larger, emergent effects depend upon the interactions of the traders.

To gain a general understanding of emergent phenomena, then, it's necessary to describe the emergence of a system's behaviour from the non-additive interactions of its building blocks (molecules, traders, etc.). A formal counterpart of the building block concept is required if the description is to encompass all complex systems. The *generators* discussed in the previous chapter provide that counterpart. Briefly said, emergence occurs when the generators for a generated system combine to yield objects having properties not obtained by summing properties of the individual generators.

10. Generating faces by recombination

To find the generators for a complex object, say a face, one typically selects features that characterize the object, such as, for a face, hair style, forehead shape, eye colour, nose shape, etc. (see Figure 10). Different faces will have different alternatives for these features, and these alternatives can be treated as Lego-like building blocks that can be put together to yield different faces.

If there are ten alternatives for each of ten features, then there are 10x10 = 100 building blocks. By choosing one building block from each feature (in some fixed order), ten billion distinct combinations can be generated. In the case of faces, that yields ten billion different recognizable faces—enough faces to yield a distinct description for each living human!

Two points are worth emphasizing:

1. The building blocks must be put together in the right order. Placing the mouth above the nose does not yield a legitimate face. The *operators* in a generated system determine which combinations of generators are legitimate, just as grammatical rules determine legitimate sentences in a language.

2. The building blocks must be carefully selected to discriminate among the objects of interest. Humans early on acquire the right building blocks for recognizing faces, but they rarely acquire the right building blocks for recognizing, say, the faces of chimpanzees. As a result, to most humans, all chimpanzee faces look pretty much alike. Similar comments apply to linguistic utterances in different languages.

With these points in mind, it's helpful to more closely examine the relations between grammars, theories of physical systems, and generated systems. Grammatical rules determine the meaningful orderings of words within a language, thereby defining the corpus for the language. Similarly, the mechanisms of a physical model (anything from levers to electron spin) determine possible trajectories through physical-state space (such as the trajectory of a probe through the solar system). It is possible to mimic grammatical rules and physical mechanisms in a generated system by specifying appropriate operators for the system. Once the appropriate operators are chosen, we can make precise comparisons between corresponding grammars, physical models, and generated systems. The generated system format offers an additional advantage because it encompasses additional important complex systems, such as computer programming languages. An important advantage of precise comparisons is that activities that are easy to observe in one complex system often suggest 'where to look' in other complex systems where the activities are difficult to observe.

The role of hierarchy in generated systems is illustrated by looking at the structures that can be constructed from a typical set of children's building blocks. From copies of a few elementary block forms (six or seven basic forms suffice), one can use a few blocks to construct a variety of simple structures: arches, towers, and the like (Figure 11). These simple structures, chosen from the multitude of possible structures, can then serve as building blocks for still more complicated structures. Herb Simon's watchmaker parable enriches this view:

> The watches the men made consisted of about 1,000 parts each. Tempus has so constructed his that if he had one partly assembled and had to put it down … it immediately fell to pieces and had to be reassembled from the elements. … The watches Hora made … [were designed] so that he could put together subassemblies of about ten elements each. Ten of these subassemblies, again, could be put together into a larger subassembly. … Hence, when Hora had to put down a partly assembled watch … he only lost a small part of his work …

With this parable in mind, contrast two descriptions of a toy castle constructed from building blocks:

1. The castle is described by listing the locations (relative to each other) of all the individual blocks used to build the castle.
2. The castle is described by first listing the locations of its larger components (walls, arches, towers), then each component is described by listing the locations of the blocks used to build that component.

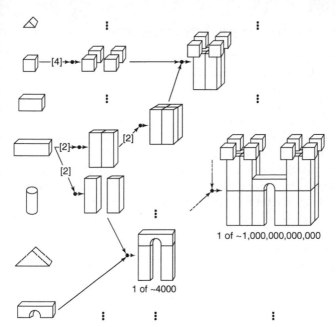

1 of ~1,000,000,000,000

1 of ~4000

11. Substructures and structures using children's building blocks

It is easy to verify that description (2) is much shorter if the larger components are used in many places. From the generated system point of view, shorter descriptions, as with shorter proofs in axiom systems, are generated earlier in the process. Early appearance in the generating process offers a modus operandi for the emergence of hierarchy. Any early description that is designated as valuable under a credit-assignment process may be used as the basis for generating further descriptions, just as the Pythagorean theorem serves as the basis for further important theorems in Euclidean geometry. Other theorems that are easy to prove but are of little use are simply ignored. Using early useful structures as a foundation for generating other structures biases the generated system to still further layered descriptions, suggesting why hierarchical organization is pervasive in complex systems.

Building blocks mediate emergence

What, then, constitutes an emergent object in a generated system? The human face emerges as a larger entity defined by the corpus of possibilities arising from allowed combinations of feature-oriented building blocks serving as generators. Much the same can be said for the English language, treating it as the entity emergent from the corpus of utterances defined by English grammar. Still, this use of the term 'emergence' may seem subjective compared to, say, 'wetness' as an emergent property of a large aggregate of H_2O molecules. To resolve this issue, it helps to look again at the concept of non-linearity. For many common properties—volume, weight, etc.—the whole *is* equal to the sum of the parts. For example, we simply sum the volumes of the various compartments of an airplane to get the total volume that must be pressurized. Much of the mathematics we use to model systems is based on this linearity—the ability to add up properties of the parts to get the corresponding properties of the whole. However, as we've seen, when the interactions between the components are non-linear (not additive), we cannot predict the actions of aggregates by simple summations. The fluctuations of equities markets ('bubbles' and 'crashes') offer an all too familiar example of emergent phenomena resulting from conditional, non-linear interactions.

Co-evolution is one of the major mechanisms for generating non-linear interactions between complex adaptive systems (CAS) agents. 'Arms races', be they between countries or organisms, provide a simple example. In a typical biological example, a leafy bush has a wide range of herbivorous insect predators. Then, as the bush evolves, it develops a protein, say quinine, that is poisonous to most insects. However, after further evolution, some insect species develops an enzyme that digests quinine. Still later, the bush evolves quinine-b that is poisonous to these insects, and so it goes on. Note that, while the bush and the insect seem to be running in a Lewis Carroll's 'Red Queen's race', neither gaining much ground with respect to the other, the co-evolving pair is better off in relation to its surroundings. The bush only has to 'protect itself' from a single predator, while the predator has a food supply that it does not have to share with other competitor species. Analogous situations occur between competing suppliers for production lines, arms races between nations, and other sets of co-evolving CAS agents. As discussed in the previous chapter, the destination of a signal is usually determined by a small part of the signal—a tag—which serves much like the address on an Internet message. For example, the leafy bush just mentioned gives off scents serving as tagged signals that attract both pollinators and predators. Quinine itself has a distinctive odour serving as a tag distinguishing the quinine-producing bush from its compatriots. In due evolutionary time, insects will begin to avoid bushes tagged with that odour. Still further down the evolutionary line, an insect that develops an enzyme for digesting quinine will actually be drawn to the quinine tag. And so it goes on.

In multi-agent systems, tags route signals through tangles of non-additive condition/action rules, implementing both stabilizing negative feedback and the amplifying effects of positive feedback. Co-evolution is often mediated by tags. Even small changes in tags can cause substantial changes in the routing of signals. The effects are quite similar to small changes in the instructions in a computer program, particularly changes in the address part of an instruction. And, just as small changes in computer programs usually cause deleterious effects, so it is that most small changes in tags cause deleterious misrouting of signals. However, in multi-agent systems, 'parent' agents remain in the system, so the damaged routing will often have little effect (other than draining off some resources) unless it actively interferes with the original routing. Thus, as pointed out at the end of the previous chapter, a complex system can explore alternatives while exploiting what it already finds useful.

Tags both serve as building blocks and can themselves be constructed from other building blocks. New routings and new processing procedures, such as production lines, can thus be generated by recombination of tags. Indeed, tag recombination provides a general mechanism for emergence, because signal-processing, in one interpretation or another, lies at the heart of all complex systems.

Boundaries and emergence

A new boundary typically identifies a new agent, so boundaries are closely associated with emergent phenomena. The semi-permeable boundaries discussed in Chapter 3 have a particularly important role. When a semi-permeable boundary encloses some volume, such as an organelle, it passes only selected reactants into that volume and restricts the exit of some products of the reaction. When the gates of an agent-defining boundary are sufficiently restrictive, the result is an agent that is a 'specialist' in processing the reactants admitted. When the products that exit one specialist agent are admitted by another specialist agent we have multi-stage processing similar to a production line. Specialists act much like catalysts: the increased concentration of reactants results in an increased reaction rate and, in a multi-stage process, the overall throughput can increase dramatically, as in Adam Smith's example.

Gated urn models (Chapter 3) allow us to reproduce and observe this 'catalytic effect' in a wide range of complex systems. Tags again come to the fore, both in specifying the gate conditions and in marking the signals or resources passed by the gates. The movement of signals between adjacent agents can often be treated as a kind of constrained diffusion: a ball is selected at random from the content of an urn, and another nearby urn is selected as a destination; if the ball satisfies the respective exit and entry conditions, then it is moved to the destination, otherwise it stays in the original urn. Repetition of this procedure, then, yields the distribution of resources (signals) when semi-permeable membranes constrain the process.

Just how is this formation of boundaries related to emergence? Because new agents serve as building blocks from which to construct new interactions and new levels of a hierarchy, they act as additions to the 'vocabulary' of the underlying 'grammar'. The operators of this 'grammar' are primarily operations, like recombination, that use extant tags as 'parents' to provide new 'offspring' tags (as shown in Figure 8). Whether the result of the new boundary is a new insect in an ecosystem, a newly distinguished pattern in a complex game, or a new cell assembly in the central nervous system, new possibilities and complexities emerge. New gates break old, rigid symmetries, resulting in emergent diversity.

When the bounded agents are situated in geometrical space, as are the reactants in Turing's reaction/diffusion model of morphogenesis (Chapter 2), the gated urn model allows us to examine the dynamics of symmetry-breaking and the emergence of patterns. It is encouraging that these complex effects can be so modelled because gated urns are readily used to model other complex systems, and because there is a close connection between gated urns and the mathematics of Markov processes (see Chapters 3 and 7 for an exposition of Markov processes). For example, questions about the emergence and persistence of distinct languages and dialects fit this format. It is plausible that dialects induce boundaries of trust, making a distinction between 'us' and 'them'. This, of course, depends upon the agents' rules and how, in an almost geographic sense, they form groups. Observation does show that groups a short distance apart along the same river, as in southeast China, can have mutually unintelligible dialects that persist over centuries. By using tags related to grammatical markers, experimental simulations using gated urn models can examine mechanisms that encourage the emergence of 'us' and 'them' boundaries. Such simulations would, in turn, suggest observational tests for these mechanisms in regions where adjacent distinct dialects are preserved. Simulations based on gated urns can be used similarly to extract underlying mechanisms in other CAS.

This formation of boundaries that mediate interactions between agents leads directly to the notion of 'niche'. Niche is a widely used term, as in 'market niche' or 'ecological niche', but it is difficult to define. The next chapter aims to put the niche concept on a firmer footing in the context of complex systems.

Chapter 7
Co-evolution and the formation of niches

What is a niche?

The term 'niche' is widely used to describe an important part of the hierarchical organization of complex adaptive systems: local use of signals and resources (as discussed in Simon Levin's *Fragile Dominion*). For instance, in ecosystem studies, the investigation typically centres on the interaction of a designated keystone species with other selected species in its locale (predators, prey, symbionts, etc.). The famous lynx/hare interactions, based on Hudson's Bay Company records, provide a specific example (Figure 12). For over a century, Hudson's Bay has kept records of the number of lynx and arctic hare pelts brought in each year. The lynx is a top predator and the arctic hare is its major prey. There are regular year-to-year oscillations in the pelt supply that can be well-described by the following factors:

The hare population is affected by the balance between an intrinsic (fixed) birth rate b_H of individuals in the population, and a death rate mainly determined by the number of lynx in the vicinity.

1. The birth rate of the lynx is proportional to the number of hares in the vicinity (its main food supply), and it has intrinsic death rate d_L from disease and old age.

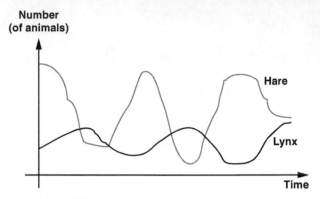

12. Lynx/Hare oscillation

Let $H(t)$ be the number of hares at time t, and let p be the proportion of the population of hares that is killed by a given lynx, then, if there are $L(t)$ lynx at time t, the total number of hares killed will be $pH(t)L(t)$. The population $H(t+1)$ of hares at the next time $t+1$ will be the population $H(t)$ at time t plus the births minus the deaths,

$$1. \quad H(t + 1) = H(t) + b_H H(t) - pH(t)L(t).$$

Similarly, the lynx population at time $t+1$ will be the population at time t plus births (proportional to the number of prey killed) minus deaths,

$$2. \quad L(t + 1) = L(t) + pH(t)L(t) - d_L L(t).$$

Given $H(t)$, $L(t)$, along with the parameters b_H, d_L, and p, this pair of equations determines the size of the next generation; then the generation at $t+2$ can be determined by substituting $H(t+1)$ and $L(t+1)$ for $H(t)$ and $L(t)$ in the equations; and so on. Thus, for given initial populations, $H(0)$ and $L(0)$, the population variation is determined from then onward, and the variations in the two populations can be plotted over time. Not too surprisingly, with

60

this strongly coupled two-agent interaction, the results closely match the Hudson's Bay data.

Equations (1) and (2) are a discrete (as contrasted to continuous) version of the famous Lotka-Volterra equations, used as the basis for a wide range of ecological niche studies. However, complex interactions, and surprises, occur even when just a few species are involved: the recent penetration of the coyote beyond its home in the western North American plains into the regions of Canada occupied by the arctic hare provides a case in point. The coyote quickly reduces the number of hares, which has caused a precipitous decline in the lynx, even though the coyote is not a predator of lynx.

Though the Lotka-Volterra equations, and their generalizations, can analyse the interactions of selected species, the associated niche concept is quite limited. Most importantly, equations like the Lotka-Volterra equations are built on the 'who eats whom' model of an ecosystem, giving little access to questions about recirculation of resources. Yet, in many complex systems, including most ecosystems, we are interested in the flow of signals and resources through a complex network of interactions that includes substantial recirculation. If we trace carbon flow in an ecosystem, we soon discover that over 90 per cent of the bio-carbon (carbon held by organisms) resides in bacteria. If we don't take account of interchanges with bacteria in broader ecosystem models, say those concerned with climate, it is as if we were trying to run a large company without knowing what is happening to 90 per cent of the cash flow!

We now know that the human body (and other eukaryotes), under normal conditions, serves as a niche for hundreds of species of benign bacteria. Indeed, there are dozens of individual bacterial cells residing in the body for each single body cell! The body cells set the structure of the body, much like the trees of a rainforest, while most interactions take place between the occupants.

This *micro-biome* is, in other words, a complex ecosystem that mediates the normal functioning of the body. When this cooperative network is unbalanced (through, say, overapplication of antibiotics) the body becomes a target for rapidly growing pathogens. The pathogens are normally kept under control by a shortage of resources—resources consumed by the benign bacteria. The benign bacteria themselves survive because the resources they consume enable the body-niche to survive over evolutionary time-scales. They have co-evolved with the body to greatly enhance the overall resilience of the whole system, through cross-species exchange of genes (*horizontal transfer*)—some genes of bacterial origin even appear in the human chromosomes. In short, the human body is a complex ecosystem that depends upon the cooperation of a vast array of bacteria for its survival.

Though the niche concept is a concept like 'complexity' that resists a sharp definition, the basic idea is that you have a diverse array of agents (organisms, market participants, industries...) that depend upon each other through an exchange of resources and signals. The recirculation of bio-carbon through bacteria in ecosystems clearly illustrates the kinds of networks associated with complex adaptive systems: networks having subnetworks with high local recirculation—the *communities* discussed in Chapter 4. If we examine flows and recirculation in a biological cell we see that membranes distinguish the different subnetworks. These membranes admit some signals and resources while rejecting others, as described in Chapter 3. The gated urns used there let us extend these ideas to niches in general. Accordingly, the tags discussed in the previous chapter play a key role in the emergence of new patterns of recirculation and, thereby, in the emergence of new niches.

Niche theory

Chapter 3 used a list of signal-processing rules to define the behaviour of a complex adaptive system (CAS) agent.

Rule 1 IF (signals w and x are present) THEN (send signals y and z)

Rule 2 IF (signals a and b are present) THEN (send signals c and d)

...   .

It was noted there that signal-processing so defined is of the same form as a list of chemical reactions

$$w + x \Leftrightarrow y + z$$

$$a + b \Leftrightarrow c + d$$

...

Using this similarity, Chapter 3 went on to introduce 'billiard ball' chemistry in gated urns as a way of analysing constrained agent interactions. The signals and rules defining an agent were treated as balls, moving on a frictionless table and interacting whenever they collided. (Billiard ball chemistry is a good first approximation to the way proteins interact in fluid media.) The concentration of each reactant in the urn can be interpreted as a probability that a random draw from the urn will yield an example of the resource. The gated urns provide a semi-permeable enclosure for the balls, limiting the exchanges between agents. With this interpretation we were able to go beyond the inadequacies of linear averages and trends for describing CAS (Figure 13).

Now we face similar problems with community-based niches. The billiard ball/urn approach can be applied to niches if we generalize the underlying reaction/encounter concept slightly. First, we list the number of members of each species in the niche. Dividing these numbers by the total number of individuals in the niche gives a set of 'concentrations' corresponding to the concentrations of reactants in the billiard ball/urn model, In other words, we can describe the niche at a given time t by describing the proportional concentration of each species within the niche at time t.

$$D(t) = <p(a,t),\ p(b,t), \ldots p(n,t)>,$$

Each *site* (square in the array) contains a set of urns representing the reactant type present in that area.

The number of balls in each urn gives the local concentration of that reactant type.

Each reactant species(same active sites) is assigned a distinct colour.

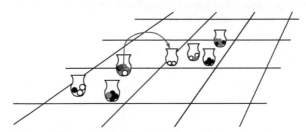

Diffusion takes place by moving balls at random between urns.

13. Diffusion between urns

Complexity

where $\{a, b, \ldots, n\}$ are the species labels, $p(a,t)$ is the proportion of species a at time t, and so on.

In algebraic terms, $D(t)$ is a vector with one entry for each species (ball colour). In the simple case of the lynx and hare oscillation we look at the total population and then observe the changing proportions over time. The proportions at any time t are represented by $D(t)$ which gives the changing probability of finding a lynx or a hare in the overall population. Because the entries are probabilities (summing to 1) the vector $D(t)$ defines a *probability distribution* (cf. the *normal distribution* or 'bell curve' that emerges in polling and sampling). The proportions, and hence the distribution $D(t)$, change as the interactions take place. Because $D(t)$ is a vector that can have many entries, we readily go to CAS having many different kinds of agents by giving the probability of each kind of agent in the system at each time t. Harking back to earlier discussions (Chapter 1), *the distribution* D(t) *specifies the state of the niche* at time t.

At this point we encounter the major difference between niches and the deterministic CAS discussed earlier: identical encounters in a niche may not always have the same outcome (e.g., the agents involved may decide to 'fight', or 'flee', or 'mate'). To allow for this possibility, we adopt from elementary probability the concept of *conditional probability* $P(h \mid (w,x))$, the probability of outcome h when the event (w,x) occurs. In present terms, $P(h \mid (w,x))$ gives the probability of outcome h when agent w encounters agent x. Thus, for each possible outcome r of the encounter we can assign a probability $P(r \mid (w,x))$, with $P(r \mid (w,x)) = 0$ if r is *not* a possible outcome of the encounter. In the niche, then, the counterpart of a CAS interaction rule is obtained by displaying a row of conditional probabilities, with one entry for each possible outcome $\{a, b, \dots, n\}$:

$$<P(a \mid (w,x)) \; P(b \mid (w,x)) \dots \qquad P(n \mid (w,x))>.$$

That is, there are n entries for the n distinct species (ball colours) in the niche; in CAS terms this row defines a signal-processing rule that allows different outcomes (signals) with different probabilities. To define the results of *all possible encounters* in the niche, we simply list the rows for all the encounters, much as we listed the signal-processing rules for an agent. Such an array of rows is, formally, a matrix M (with the same properties as matrices in elementary algebra) and, because the entries are conditional probabilities, it is called a *Markov matrix*. The matrix M provides a solid base for niche theory because matrix M and vector $D(t)$ can be multiplied much like numbers (though the details are not so simple) to give the change in $D(t)$ caused by an encounter in the niche.

Recall that in a billiard ball chemistry a particular encounter is a random event, depending on which reactants happen to collide. And, in a niche, the encounter itself may have different outcomes. These combined effects change the distribution $D(t)$. Multiplying $D(t)$ by M in fact gives the modified distribution $D(t+1)$ resulting from the possible encounters (i.e., $D(t+1) = D(t)M$). The process can be repeated, so $D(t+2) = D(t+1)M = D(t)M^2$, or in general $D(t+T) = M^T$. In other words, if we want to explore the result of a sequence of interactions in the niche over T units of time, we use M^T.

Though the theory of Markov processes is too intricate to describe here, William Feller's well-known text *Probability Theory and Its Applications* provides an easy exposition with suggestive examples. The net result of this journey into Markov processes is a mathematical theory of niches that allows for multiple species with interaction networks that involve loops and recirculation. Once the Markov matrix M is defined for a niche we can investigate dynamics that go well beyond the 'who eats whom' paradigm exemplified by the lynx/hare oscillations described at the beginning of this chapter. In particular, we can investigate the dynamics of niches like the cup-like bromeliad of the tropical rainforest (Chapter 1). That niche involves a complicated network involving the plant, insects, and amphibians, mediated by retained rainfall. Such interactions are not well described by averages and trends. Similarly, a 'market niche' depends upon a network of suppliers and customers that is constantly shifting, where departures from averages and trends, such as 'bubbles' and 'crashes', are common. So it is with other community-based niches.

The main advance provided by Markov processes is a theory that describes the complicated recirculation (feedback, both positive and negative) that occurs in realistic community-based niches. The number of different ways such a niche can be occupied is huge. Even if there are only two species (colours) with a total of only five individuals (balls) placed in three locations (urns), there are 40,320 distinct arrangements (as can be determined by elementary, but tedious, calculations that I will not reproduce here) (Figure 14). However, from any given arrangement, only relatively few other arrangements can be reached in one step—at most 12 arrangements in the case of five balls, two colours, and three urns (try it). Thus, the corresponding Markov matrix M has mostly zero entries in a given row—only 12 non-zero entries in the

A$_h$, one of 40,320 distinct arrangements of 5 balls (individuals) of 2 colours (species) in 3 urns (locations)

Moving 1 ball to change the arrangement A$_h$

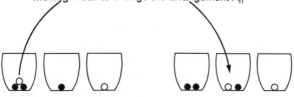

9 arrangements can be obtained from A$_h$ by moving 1 ball:

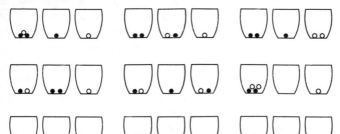

14. Ball arrangements in urns

example just given. This *sparseness* of M simplifies both the computational and theoretical analysis of the corresponding niche.

Niche formation through co-evolution

The definition of a niche as a set of conditional (probabilistic) rules—the rows of the Markov matrix—provides an avenue for investigating the ways in which niches form. When we look at realistic niches, whether they be market niches or niches in a rainforest, the *diversity* of the niche dwellers stands out. Unlike the simple 'who eats whom' models, we see a complicated recirculation of resources and signals—think of the bromeliad (Chapter 1) or the micro-biome discussed at the beginning of this chapter. How did this complex network of interactions evolve?

The short answer is *co-evolution* through recombination of building blocks (Chapter 6), often accompanied by an exaggeration of some of the resulting characteristics (as in the hawk moth and comet orchid discussed in Chapter 5). Cascades of increasingly specialized agents result. As nicely described by Samuelson in his classic text *Economics* there is a *multiplier effect* in cascades: in everyday terms, when you pay a carpenter to make changes in your house the cash does not disappear; instead the carpenter pays some of it to, say, a grocer, who in turn pays a supplier, and so on. Along the way, some of the cash *does* get diverted from the cascade into savings, taxes, and the like, but the overall effect goes well beyond your initial injection. The multiplier effect in a typical cascade may be 4 (or more), indicating that the initial payment has the effect of four separate injections of cash (without the pass-on). Similar comments apply to production lines (à la Adam Smith's 'pin factory' example), protein metabolism in an organelle, and other CAS niches.

The multiplier effect that accompanies the re-use of resources in a cascade typically drives the occupants of a niche to increasing specialization. Darwin's comet orchid illustrates the extremes of specialization that can result. Similar changes over time can be observed in most niches: transportation niches in an economy, dialects and slangs in languages, special trading procedures in markets (e.g. 'derivatives'), and so on. When a new variation arises in a niche, it usually makes possible new paths in the network of interactions. If a new path provides an increase in the multiplier effect (often through increased recirculation) it persists. The long-term effect is increasing diversity that breaks initial uniformity—Turing's reaction/diffusion model of morphogenesis (Chapter 2) writ large.

Whole niches can undergo major modifications and even replacement. Replacement typically occurs when the niche's raison d'être is taken over by a new niche. For example, the transportation niche used to produce all the components for horse buggies, including buggy whips, but the advent of automobiles gave rise to a whole new set of producers, with accelerator components replacing buggy whips. Similarly, the replacement of large dinosaur carnivores by large mammalian carnivores illustrates a major niche shift on a geological time-scale. In this shift, one community has been replaced by another that has similar internal connections but involves different nodes (species). When agents are modelled by gated urns, the shift is marked by a different set of tags (with a different co-evolutionary history) yielding a similar network of interactions.

The multi-agent nature of niches, and CAS in general, makes it possible to explore the relevance of large numbers of specializations while exploiting what is already useful. However, even for small systems the number of possibilities is so enormous that exhaustive search is not feasible. Recall the faces constructed from building blocks (Chapter 6): ten billion faces can be generated using only 100 components. In a gated urn model, a gate sensitive to binary tags at ten specified loci can select amongst $3^{10} = 59,049$ tags (see Chapter 4). If there are urns with five gates attending to tags at five non-overlapping ten-bit loci, then there are $(3^{10})^5 = 3^{50} > 10^{23}$ different routings. That is an astronomical number, but it is still small compared to, say, the possible interaction networks for biological cells or systems of neurons. So, despite the ability of a multi-agent system to explore alternatives simultaneously, there is no hope of exhaustive testing.

There is, however, a different mode of exploration. It proceeds by constructing network communities in a hierarchical fashion. Instead of a single 'who-eats-whom' community, communities are used as building blocks to form larger communities. That is, tightly connected communities exchange some resources with other communities to form larger, less tightly connected, communities. By repeating this process, it is relatively easy to form a hierarchical network with many levels. The resulting combined niches are similar to combined subroutines in a computer program, performing ever more complex transformations of signals and resources. As in the design of computer programs, the hierarchical (modular) approach makes it feasible to search out complex transformations that would be rarely uncovered in a program-by-program exhaustive search. Of course, the building block communities at any level are only a small selection of the possible communities at that level. But they can be tested in feasible times, a point eloquently made by Herbert Simon (discussed in Chapter 1). Feasible testing times account for the fact that hierarchical organization is a pervasive feature of all complex adaptive systems.

Is it possible to find *mechanisms* for generating hierarchies that are common to all CAS?

Gated urns, and the related Markov matrices, offer one path. Chapter 6 pointed out the role of *recombination* in generating new gate conditions. Because limited parts of the signals, the *tags*, almost always determine passage through gates, interest centres on the effects of recombination on tag-based conditions. If we examine the recombination of two tag-based gate conditions, the resulting pair of 'offspring' conditions usually consists of one condition that will accept a broader range of tags and another that accepts a smaller range (see Figure 8). The changes may be small, but they can sometimes cause quite large changes in the distribution of balls (species) within the urns (niche). A few pencil-and-paper experiments with small models clearly show this effect.

The discussion of Markov models of gated urns at the beginning of this chapter pointed out that relatively few re-arrangements of balls can be achieved by moving a single ball from one urn to another. That means that the network corresponding to the urn model (where nodes represent arrangements) has low 'fanout' (degree)—each node connects to relatively few other nodes. In such a network, co-evolutionary changes resulting from changed tags stand out because they are so localized.

The net effect of a succession of changes in tags is a series of snapshots of the changing network, producing a kind of 'movie' of co-evolutionary changes. If the snapshots do not change too rapidly, the flows over the network will 'settle down' to something approximating a *steady state* distribution of balls in the urns. From a technical point of view, steady state distributions are *fixed points* of the Markov matrix M defined by the urns (this is clearly shown in Feller's *Probability and Its Applications*). When the steady state is known, it is easy to determine critical properties of the niche, such as its throughput (Chapter 5). For instance, if the throughput increases under a newer tag-defined routing, that routing will be retained, typically increasing the diversity in the tagged urns that define the niche.

At this point, connections have been made between three different ways of presenting a niche:

- Networks provide a precise snapshot of the interactions of agents in a complex system at a given point in time

- The communities in these networks define existing niches within the complex system

- For each community there is a corresponding set of gated urns, and recombination of the tag-sensitive gate conditions provides a precise approach to co-evolutionary succession within the complex system.

The Markov matrices corresponding to the gated urns, as discussed in the first section of this chapter, open an avenue for mathematical analysis. Under this formulation, the study of niche formation is the study of changes in recirculation as agents in the niche co-evolve—or, more formally expressed, changes in the fixed points of the Markov matrix.

Mechanisms producing niche evolution

Now we are ready to more closely examine the mechanisms that can generate and modify hierarchies. As discussed earlier, in the previous section and in Chapter 5, the mechanism of tag recombination provides a natural way to generate new networks from extant networks. Recall that recombination typically produces one offspring that has more specific requirements than either parent, while the other offspring has less restrictive requirements (see Figure 8). A more specific tag requirement may link a pair of agents so that they act in sequence, accomplishing the same task as some extant generalist. As in Adam Smith's example of pin production, the throughput associated with the linked pair will usually increase, contributing to the survival of the linked agents. On the other hand, a more general gate requirement may accept signals belonging to other production lines, thereby increasing the input to the niche that contains the parent agents.

Viewed this way, mechanisms for recombining tags can be looked upon as mechanisms for niche construction and co-evolution. Tags produced by recombination use tested building blocks (parts of tags already proved useful) to produce an extraordinary variety of routings within the niche. As mentioned earlier, new routings, because of the parallelism of CAS, act alongside the routings provided by the parents, so the new routings can usually be explored without abandoning capacities already attained.

The critical question then becomes: once new possibilities have been generated, how does a multi-agent system select amongst the various organizations so produced? As a starting point, consider two species for which all resources required for replication are freely available, with one exception, which they both require and which is in short supply. In this case, the species that uses that resource more efficiently will rapidly increase its proportion in the population relative to the other species. As its numbers increase it will take more and more of the critical resource, 'starving' its competitor. More generally, and almost by definition, organizations that make more efficient use of limited resources required for existence are the ones that survive (a version of Herbert Spencer's 'survival of the fittest'). Typically, this increased efficiency will be the result of production-line-like processes, such as coupled reactions, recirculation, and autocatalysis, created by successive applications of recombination.

Looked at from the network point of view, survival depends on routings that provide better use of resources—for example, recirculation of water and bio-carbon in an ecosystem, or an improved multiplier effect for cash flow in an economy. Inside a network community, it is these routings that are favoured. As a consequence, the tags that determine these routings become persistent features of the community. Persistent tags, in turn, become building blocks for further changes, and, consequently, a source of new hierarchical organizations. In this process, the throughput from lower level niches largely determines the nature of interactions between higher level niches—persistent lower level niches with higher throughput have more influence on the overall organization.

From the credit-assignment point of view (Chapter 3), a rule that accepts a signal it did not previously accept becomes a new buyer for the rule that generated that signal (the seller). The new buyer thus increases the income of the seller. The new buyer–seller link may 'set the stage' for later acquisition of valuable resources, with such middleman setting the stage for the next in line. In this linked chain the critical resource is the 'payment' made to the ltimate consumer', the last in line. For example, this 'payment' could be a resource critical to the survival of the whole chain— the chain of middlemen leading from ore mining to automobile production provides a familiar example. Credit assignment, in this format, ultimately determines the survival of the linked agents within their community.

It's worth emphasizing that the mechanisms for niche formation discussed here are not the only ones, nor are they necessarily canonical, but they are pervasive and they emphasize the importance of mechanisms that generate hierarchical boundary arrangements.

Chapter 8
Putting it all together

The role of an overarching framework

Because the study of complex systems, and complex adaptive systems in particular, involves a variety of models and concepts, a framework that uses a common language to describe these systems offers a substantial advantage in making comparisons. Chapter 1 emphasized the importance of *generated systems* for understanding complexity, using games, Euclidean geometry, and Newton's laws as examples. Chapter 6 added emphasis to the role of generated systems by pointing up their relation to the phenomenon of emergence. A generated system, then, offers a way of constructing an overarching framework that makes comparisons possible: it describes the pieces of different complex systems in a single language, while emphasizing critical common properties such as emergence.

As described in Chapter 6, a *generated system* consists of a set of *generators* (essentially an alphabet) and a set of *operators* (a set of rules for combining letters from the alphabet to form legitimate 'sentences'). The set of combinations generated, the *corpus*, provides a precise labelling of objects of interest (theorems, sentences, DNA sequences, and so on). That it is possible to find a generated framework applicable to all complex systems is suggested by the existence of universal computing languages generated from a small

, of basic instructions. If we set up a generated system that embodies a universal computing language then any possible algorithmic model of a complex system can be described and studied therein. When the framework is well-designed, the generators serve as Lego-like building blocks (Chapter 6) from which to construct different kinds of complex adaptive systems (CAS) agents.

When it is possible to compare similarities and differences between systems, activities that are difficult to observe and understand in one complex system can often be related to activities that are easily accessible and understandable in another. Thus, a precise comparison frequently suggests new ways to observe and understand activities in the 'difficult-to-observe' situations. Our search, then, is for a universal language 'tuned' to the description of complex systems, particularly CAS.

Because the dynamics of a CAS, especially the processes involved in adaptation, are of central interest, we want a time-like assignment to the elements of the generated corpus. A straightforward way to accomplish this assignment is to let the order of generation of the elements correspond to their time of first appearance. That is, the generating procedure produces a counterpart of Darwin's 'tree of life'. Accordingly, under adaptation, the structures of a new 'generation' are generated from selected structures of the previous generation. As we noted in Chapter 6, this way of proceeding is not very different from selecting the theorem of Pythagoras in geometry as a source of new, related theorems. In CAS, it is the succession of new combinations of building blocks—the emergence of new agents—that is of interest.

A closer look at generated systems

To get a better feeling for what a generated system entails, let's start with a familiar domain, the addition of positive and negative integers. The operator in this case is the familiar '+', used to form the sum of two numbers, as in 1+1 =2. The number 0 acts as an *identity* element, in the sense that 0 added to any integer simply gives back the same integer, as in 0+9 = 9. The negative of any integer acts as its *inverse* in the sense that the sum of an integer and its inverse yields the identity element, as in –9+9 = 0. Finally, there are two technical points:

1. + is *commutative* in the sense that when integer x is added to y, the result is the same as y added to x, as in 3+2 = 2+3;

2. + is *associative* in the sense that the sum $x+y$ added to z is the same as x added to the sum $y+z$, as in (2+3)+4 = 2+(3+4) = 9.

It is worth noting that there are 'number systems' that do *not* satisfy these last two technical requirements, just as there are non-Euclidean geometries.

When the + operator satisfies these requirements, the additive arithmetic of the entire set of positive and negative integers can be *generated* from an initial set of just three generators, {–1, 0, 1}, using the operator +. That is, by adding 1 to itself n times we get the integer n, and any + operation on two integers yields another integer. Mathematicians have generalized this idea to the concept of a *finitely generated group* which is specified by an operator x and a set of generators

$$\{e, g_1, g_2, \ldots, g_k, g_1^{-1}, g_2^{-1}, \ldots g_k^{-1}\},$$

where g_j^{-1} is the inverse of g_j and e is the identity element $g_j \times g_j^{-1} = e$. Finitely generated groups have been studied for over a century and it is an interesting sidelight that there are still important open questions about them. We should expect no less for a finitely generated framework for complex adaptive systems.

A generated framework for CAS

A short introduction is not the place to present the technical details of a formal overarching CAS framework, particularly the ighly formal details of a *finitely generated framework*. Still some of the framework's key requirements can be described here.

Any CAS should be describable therein. An easy way to accomplish this objective is to make the framework computation universal, as suggested at the beginning of this chapter, so that any model that can be simulated can be described within the framework. In practice, this means setting up a generated system wherein the operators amount to the basic 'machine-level' instructions of a general purpose computer (typically the 'microchip' for a computer provides for the execution of 32 or 64 such instructions). *Agent rules and subroutines should have direct descriptions.* The framework's operators should combine the generators provided to form agent-defining programs.

Hierarchies should be easily formulated. The organelle/cell/organ/... hierarchies of biological organisms, the nested communities of ecosystems, the dialectical organization of language communities and other CAS hierarchies should have obvious descriptions. Ease of description is important because formally equivalent systems are often not equivalent in ease of application. For example, there are axiom systems that generate exactly the same corpus of theorems as Euclidean geometry but in which the shortest proof of the Pythagorean theorem exceeds any pre-determined limit. A geometry with no feasible proof of the Pythagorean theorem will obviously differ greatly from the geometry we use.

Populations should have simple descriptions. In particular, combinations of exploration and exploitation (Chapter 5) should be easy to explore.

CAS dynamics should have an explicit representation. To accomplish this objective, successive steps in generating the corpus must correspond to time-like variations in the structure of the CAS (Chapter 7). The operators must facilitate adaptation and evolution (via mechanisms such as mutation, cross-over, and selection according to fitness).

Important questions

As is true of most systems, there are important questions about complex systems that cannot be answered by observation alone. Even Newton faced questions about physical systems that cannot be answered by simply collecting data: 'Why do objects on earth come to rest, while planets seemingly move on forever in their orbits?' A typical question about complex adaptive systems that requires more than observation centres on human populations: 'How do we remain human when no individual in the offspring generation is identical to any individual in the parent generation?' Even faces are unique, let alone deeper characteristics such as metabolic networks. Yet there must be common persistent characteristics that make us human and enable us to survive in the highly variable 'human niche'. 'What is preserved and how?' A similar question applies to all CAS, because all CAS consist of populations of adaptive, interacting agents. Indeed, the resilience of a CAS when confronted with 'shocks' (invasive species, new trade conventions, or the like) generally depends upon inherited, *persistent characteristics*.

In searching for persistent characteristics in a CAS it is helpful to think again of the effects of cross-over vis-à-vis motifs (Chapter 3). The motifs that persist through successive generations typically select for mechanisms that enable the agents to thrive (gates in semi-permeable membranes, catalysts, and the like). There are indeed theorems about *genetic algorithms* (*GAs*) which detail the persistence and spread of motifs within systems that evolve under the action of recombination. The framework provided by gated urns and Markov matrices makes it possible to compare the co-evolution of motifs in a variety of CAS niches, often suggesting new mechanisms for control.

There is another important question about CAS that can only be answered with the help of a theoretical framework: How do new, *semi-autonomous agents* arise in CAS? As discussed in Chapter 4, autonomy is closely associated with loops and recirculation in networks. As a result, the activities of semi-autonomous agents are only partially controlled by current input. So agents can examine different courses of action prior to execution, as when humans examine alternatives in their tactics in a game (Chapter 4). Occasionally the recirculation is easy to observe (e.g., tracing the flow of money in a small-town economy), but in an ecosystem it is difficult to trace the flow of bio-carbon (where 90 per cent of the bio-carbon may be processed at some stage by unknown bacteria). In tangled networks, such as the mammalian central nervous system, most neurons belong to hundreds of loops. In these cases, it is difficult to determine the changes in autonomy caused by adding even a few additional loops. Still, comparing observations in 'easy cases' (cash flow) to observations in difficult cases (the central nervous system) may suggest ways of transferring techniques for control (e.g., budgeting) from one to the other.

Another important question is posed by one of the most obvious characteristics of CAS: *diversity*. How does a CAS retain its diversity in face of continual pressures for adaptation and improvement? Partly, this diversity is the result of the extensive interactions between component agents. Because of continually varying context provided by other agents, it is difficult to define what an optimal agent could be, let alone show that it would 'take over' the system.

Observation shows that agents acting in a niche continually undergo 'improvements', without ever completely outcompeting other agents in the community. These improvements may come about in either of two ways: (i) an agent may become more of a generalist, processing resources from a wider variety of sources, or (ii) it may become more specialized, becoming more efficient than its competitors at exploiting a particular source of a vital resource. Both changes allow for still more interactions and still greater diversity. Often two species of agents become involved in a Red Queen's offence/defence race (Chapter 6), where the prey species evolves a defence (thereby cutting down the number of predators), and the predator then evolves a counter to the defence (thereby gaining a resource not open to its competitors). While, over time, neither member of this pair seems to gain a persistent advantage over the other, both are progressively better off in the larger context of other prey and predators. These highly co-evolved agents are a persistent feature of CAS, contributing to the overall diversity.

The ease of extracting and defining mechanisms for improvement varies greatly from one CAS to another. As emphasized at the beginning of this chapter, precise comparison of the varied mechanisms opens the possibility of 'transferring' mechanisms from clearly defined settings to other less clearly defined settings.

Here I'll conclude with one more question (though the list could be longer) where an answer can only be attained with the help of a theory: Why are all CAS organized in a *hierarchical* fashion? Observation shows that different levels of a CAS hierarchy are subject to different 'laws'. The laws are self-consistent, in the sense that laws at one level do not violate laws of previous, lower levels. However, the laws are not fully determined by the lower level laws. For example, the flight of a dragonfly depends on airflow around the wings, governed by the Navier- Stokes equations, but the overall control of the flight path depends upon an entirely different set of equations.

For some systems, the laws applying to a given level are relatively easy to derive from observations, but they can be quite obscure in other systems. Once again, when precise comparisons can be made, it is often possible to infer the form of laws that give an advantage to hierarchical organization in all CAS. As in the case of the dragonfly, the laws at various levels of the hierarchy are closely linked to the possibilities for 'steering' complex systems.

Interpretation of the framework

The framework outlined here is quite grammar-like, but with the intention of offering different interpretations in different contexts. For example, the same string of generators may be interpreted as a protein in one context and as a message on the Internet in another. Such 'ambiguity' is deliberate—the framework is meant to supply a common rigorous language for describing a wide variety of CAS. A particular interpretation stems from assigning an interpretation to each individual generator, much as one assigns a mass and velocity to each particle in Newtonian framework. Thus, in the protein case the generators are taken to be amino acids, while in the Internet case they are interpreted as bits. For CAS, it is important that the interpretation assigned to the generators should facilitate the interpretation of short strings of generators as building blocks for higher levels of organization. Indeed the framework should facilitate the use of selected combinations of generators at one level as generators for an embedded higher level grammar—the counterpart of forming new laws at higher levels in physics and chemistry (Chapter 1).

Once the framework's generators and operators are interpreted to model a specific CAS, it is natural to ask whether or not the model is a good model of that system. Here the interplay between the model and observations of the system comes to the fore. The particular CAS of interest is a selected part of the broader world within which it is embedded (roughly, the environment of the CAS). For present purposes we can think of this broader world as law-governed, but according to unknown laws. Moreover, think of the world as being *perpetually novel*, so that identical situations rarely, if ever, recur. Even the 'small world' provided by the game of chess exemplifies this perpetual novelty—under tournament conditions, each completed play of the game is different from all other plays in the recorded history of the game. Under the conditions of unknown laws and perpetual novelty, a model is based on recurring patterns in the world (Chapter 1).

A recurring pattern means that some *properties* of world states recur, even though the states themselves do not recur. And, of course, these properties must be observable through some set D of available *detectors*, such as the gauges or instruments in a control panel (Chapter 3). The observations provided by the detectors (fuel level, engine temperature, etc., in an automobile or airplane) provide direct information about the state of the system, though there will be many other aspects of the system that are not detected (e.g., control panels usually do not report on 'cracks in the structure'). A straightforward way to build the model is to assign a state of the model to each set of readings. Said another way, all possible readings for the detectors (the corpus of readings) are assigned (as interpretations) to different combinations of generators in the framework. Once this is done, the operators of the framework also transform the model state for one time-step into the model state one time-step later (see Figure 15)—that is, the operators supply the 'laws' of the model.

A model is a good model when:

> A time-step in the model corresponds to a specified elapsed time in the world and
>
> The model's laws transform its state so that resulting state matches the detector readings after the specified time has elapsed in the world.

In short, the model *predicts* the new readings expected from a later observation. For example, if the model's time-step corresponds to 24 hours, and the observations are of a planet's position in the night sky, then the model (e.g. the ellipses of

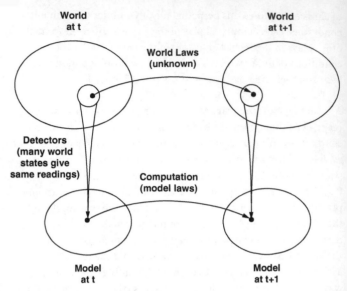

15. Criterion for a valid model

Copernicus) predicts the planet's position tomorrow from the observed position today.

Mathematically, the requirements imposed by (1) and (2) are requirements that the map of detector readings of the world W to states of the model M, combined with the laws of the model, form a *commutative diagram*: going forward one time-step in W then down via the detectors to the model state ('forward'/'down') should yield the same result as going down via the detectors then forward via the model laws ('down'/'forward') (see Figure 15). In short,

$$('forward'/'down') = ('down'/'forward').$$

When this relationship is formally defined in mathematical terms it is called a *homomorphism*.

A short summary

Emergence is the primary characteristic that distinguishes complex systems as an important subset of complicated systems. This distinction is sometimes difficult in practice because there is a continuum of borderline cases that might or might not be classified as examples of emergence. Nevertheless, just as there are clear cases where a collection of sand grains constitutes a 'pile', there are undisputed examples of emergence. These examples range from 'wetness' as an emergent property of aggregates of water molecules to 'self-reproduction' as an emergent property of sequences of 'instructions' (be they patterns in a cellular automaton or strings of RNA and DNA).

Emergence is tightly tied to the formation of boundaries. These boundaries can arise from symmetry breaking, as in Ising's bar magnet model or Turing's model of morphogenesis (Chapter 2), or they can arise by assembly of component building blocks, as in the assembly of semi-permeable membranes from proteins (Chapter 3). For CAS, the agent-defining boundaries determine the interactions between agents. Indeed the agent-defining boundaries act like semi-permeable membranes, admitting some signals and excluding others. Adaptation, and the emergence of new kinds of agents, then arises from changes in the relevant boundaries.

Typically, a boundary only looks to a small segment of a signal, a *tag*, to determine whether or not the signal can pass through the boundary. For proteins the tags are active sites and ligands, for the internet tags are message headers, for ecosystems tags are species phenotypic characteristics (such as shape, colour, scent, and the like), and tags are similarly distinguished in other complex systems. A boundary sorts signals according to tags in a rule-like way:

IF (signal x with required *tag r* is present) THEN (pass signal x through boundary).

An agent processes signals in a similar way:

IF (signal x with required *tag* r is present) THEN (emit signal y).

Often the IF part contains more than one condition and the THEN part may emit more than one signal, as when two chemicals react to produce two new chemicals. So an agent can be modelled by a set of conditional IF/THEN rules that represent both the effects of boundaries and internal signal-processing. Because tags are short, a given signal may carry multiple tags, and the rules that process signals can require the presence of more than one tag for the processing to proceed.

Agents are *parallel processors* in the sense that all rules that are satisfied simultaneously in the agent are executed simultaneously. As a result, the interior of an agent will usually be filled with multiple signals (e.g., the interior of a biological cell contains multitudes of proteins that serve as signals). The central role of tags in routing signals through this complex interior puts emphasis on the mechanisms for tag modification as a means of adaptation. Recombination of extant conditions and signals (as discussed under 'Rule discovery' in Chapter 3 and in Chapter 5), turns tags into building blocks for specifying new routes. Parallel processing then makes it possible to test new routes so formed without seriously disrupting extant useful routes.

Sophisticated agents have another means of adaptation: anticipation ('lookahead'). If an agent has a set of rules that simulates part of its world, then it can run this internal model to examine the outcomes of different action sequences *before* those actions are executed. That is, the rules that determine this internal model act much as the laws of a board game that allow one to plan a sequence of actions leading to a desirable future board configuration. For this 'lookahead' simulation to be possible, the 'internal model' must be autonomous: the current input to the agent should only initiate activity in the internal model. Then, the rules that specify the internal model run independently of current agent input, using a sequence of internally simulated inputs to examine future possibilities. Roughly, in an autonomous subsystem, the current input only modulates the ongoing activity.

Here we come upon a substantial contrast with much of the research that is labelled 'machine learning'. Most machine-learning models, including 'artificial neural networks' and 'Bayesian networks', lack feedback cycles—they are often called 'feedforward networks' (in contrast to networks with substantial feedback). In the terms used in Chapter 4, such networks have no 'recirculation' and hence have no autonomous subsystems. Networks with substantial numbers of cycles are difficult to analyse, but a large number of cycles is the essential requirement for the autonomous internal models that make lookahead and planning possible.

The most productive way to model CAS that have internal models is to describe the CAS models within an overarching framework that makes rigorous comparisons possible. *Gated urns* (introduced in Chapter 3) offer an approach to such a framework. The gated urn format uses a modification of the classic urn model of probability theory, where the proportion of, say, black balls in the urn determines the probability of drawing a black ball. To extend this model to signal-processing, different colours of balls are used to represent different signals. Each urn, then, corresponds to a boundary that marks an agent or encloses an agent aggregate. Broadcasting a signal locally (diffusion) is modelled by repeatedly moving a randomly selected ball from one urn to others. The gates of a gated urn restrict the signals that can be broadcast (moved) by restricting the colours of balls that can enter (or leave) an urn. Thus, a gated urn acts much like a semi-permeable membrane in a biological cell, admitting or excluding signals according to the tags they carry.

A formal framework based on gated urns provides access to an important means of analysis, the theory of Markov processes (as tlined in Chapter 7). Basically, the Markov process specifies ways in which the states of a system can change over time, much like the rules of a game specify the board configurations attainable from any given configuration. For a set of gated urns, these states are simply the possible distributions of balls of various colours in the urns, and changes in the distribution are caused by signal-broadcasting (movement of balls between urns) and signal-processing (which changes the colours of balls in an urn). With a finite set of balls and urns, there will be a finite number K of possible distributions. The rules governing all possible state changes can then be specified by a $K \times K$ array, the Markov matrix: The entry for row s of the matrix at column s', designated $P(s'|s)$, is the probability that the system will go from state s to state s' in one time-step. Given this matrix, all possible sequences of state change over an interval T can be determined by standard matrix multiplication.

It is interesting, and useful for the study of CAS, that the Markov formulation emphasizes the critical role of tags: any change in a tag modifies the entries $P(s'|s)$ for every state change that involves that tag. As in physics, questions about the mechanisms that generate change are particularly important. Do similar mechanisms in different CAS operate on tags to produce a similar co-evolution of signals and boundaries? Do the interaction networks so-produced go through similar stages of evolution? And, what mechanisms make possible the increasing diversity exhibited by CAS as time elapses? When looking for general answers to these questions, the framework based on gated urns provides a way of making comparisons that go beyond rhetoric.

To see an example of a useful comparison, consider using the gated urn format to describe Adam Smith's classic example of a 'production line'. In that example, a generalist blacksmith producing straight pins (first drawing wire, then clipping it, then adding a head, etc.) is pitted against a 'production line' of specialists, each one specializing in one of the actions of the blacksmith. The gated urn format suggests a general way to represent this 'production line': each stage is represented by an urn which has a gate that admits signals from the previous stage as well as a gate that passes signals resulting from its interior processing. The throughput of this 'production line' can be compared to the throughput of a single urn that produces the same product via a set of interior reactions. As Smith points out, and as can be verified with detailed calculations using the urn model (this is laid out in Holland's *Signals and Boundaries*), the throughput of the sequence of specialist urns is many times that of the single generalist urn. If the resulting product is involved in the replication (or survival) of the agents involved, then the new arrangement quickly becomes established in the CAS, suggesting why cascades of specialists, and the concomitant diversity, are prevalent features of CAS.

Recombination abets the formation of new cascades. When gating conditions are crossed (as described in Chapter 3), one offspring gate is typically more specific in its tag requirements than either parent gate. On occasion, the more specific gate links that urn with the output of an extant urn, so that the new urn takes in and processes the extant urn's output. That is, a two-urn 'production line' results. When the throughput of the cascade is relevant to the agent's survival, it becomes a building block for longer cascades.

By extracting general mechanisms that modify CAS, such as recombination, we go from examination of particular instances to a unified study of characteristic CAS properties. The mechanisms of interest act mainly on extant substructures, using them as building blocks for more complex substructures (as described in Chapter 6). Because signals and boundaries are a pervasive feature of CAS, their modification has a central role in this adaptive process. As a result, there are good opportunities for developing a unified theory of pervasive CAS characteristics such as niches and hierarchical organization (Chapter 7).

It helps, in developing a theory, to test it against examples not used in formulating the theory. Here, viewing language as a complex adaptive system serves that purpose (Chapter 6). From the CAS point of view, languages set boundaries on communication between human agents. Within the gated urn framework, one can model and examine the plausibility of the hypothesis that a shared language, or dialect, sets boundaries of trust. That is, a dialect distinguishes between 'us' and 'them', much as troops of chimpanzees use their calls to distinguish their own troop from other troops (which they treat as invaders). As language evolves, do the mechanisms extracted from other CAS operate to provide and preserve dialects? If so, it adds plausibility to the generality of the candidate theory.

It has been the overall objective of this short introduction to give a general idea of what we know and don't know about complex systems, with an emphasis on how an overarching theory would increase our understanding. In looking at language acquisition as a test case, we see how mechanisms, and the structures they help build, tell us 'where to look' when trying to understand complex systems. It is clear that complex systems are still primarily at the stage of collecting and examining examples, much as was the case in the early stages of biology, or the early stages of physics before Newton, or the study of electrical and magnetic phenomena before Maxwell. We are still a long way from an overarching theory of complexity, but there is strong evidence that such a theory is possible.

Further reading

Chapter 1: Complex systems

Herbert A. Simon. *The Sciences of the Artificial*. MIT Press (1981)
Arthur W. Burks. *Essays in Cellular Automata*. University of Illinois
    Press (1970)
Mitchell M. Waldrop. *Complexity*. Simon & Schuster (1972)

Chapter 2: Complex physical systems (CPS)

Melanie Mitchell. *Complexity: A Guided Tour*. Oxford (2009)

Chapter 3: Complex adaptive systems (CAS)

John H. Holland. *Hidden Order*. Perseus (1995)

Chapter 4: Agents, networks, degree, and recirculation

Donald O. Hebb. *The Organization of Behavior*. Wiley (1949)

Chapter 5: Specialization and diversity

Mea Allan. *Darwin and his Flowers*. Taplinger (1977)
Adam Smith. *The Wealth of Nations*. Penguin (1982)

Chapter 6: Emergence

John H. Holland. *Signals and Boundaries*. MIT Press (2012)

Chapter 7: Co-evolution and the formation of niches

Simon A. Levin. *Fragile Dominion*. Perseus (1999)
William Feller. *Probability Theory and its Applications*. Wiley (1950)
Paul A. Samuelson. *Economics*. McGraw-Hill (1948)

Index

Index

Expand your collection of
VERY SHORT INTRODUCTIONS

MATHEMATICS
A Very Short Introduction
Timothy Gowers

The aim of this book is to explain, carefully but not technically, the differences between advanced, research-level mathematics, and the sort of mathematics we learn at school. The most fundamental differences are philosophical, and readers of this book will emerge with a clearer understanding of paradoxical-sounding concepts such as infinity, curved space, and imaginary numbers. The first few chapters are about general aspects of mathematical thought. These are followed by discussions of more specific topics, and the book closes with a chapter answering common sociological questions about the mathematical community (such as "Is it true that mathematicians burn out at the age of 25?")

www.oup.com/vsi